ISLANDS IN TIME

Island sociogeography and
Mediterranean prehistory

Mark Patton

London and New York

First published 1996
by Routledge
11 New Fetter Lane, London EC4P 4EE

Simultaneously published in the USA and Canada
by Routledge
29 West 35th Street, New York, NY 10001

Routledge is an International Thomson Publishing company I(T)P

© 1996 Mark Patton

Typeset in Garamond by
Florencetype Ltd, Stoodleigh, Devon

Printed and bound in Great Britain by
Biddles Ltd, Guildford and King's Lynn

British Library Cataloguing in Publication Data
A catalogue record for this book is available from the British Library

Library of Congress Cataloguing in Publication Data
A catalogue record for this book has been requested

ISBN 0–415–12659–2

ISLANDS IN TIME

CONTENTS

For Amanda and Caj

FIGURES

FIGURES

TABLES

ACKNOWLEDGEMENTS

The idea for this book first arose in 1984, whilst I was working in Cyprus as a site supervisor on the excavations at Kalavassos-Tenta. I had previously carried out research in the Channel Islands and, turning my attention to the archaeology of the much larger island of Cyprus, I became increasingly interested in the general question of insularity and its effect on prehistoric cultures. I am therefore grateful to Ian Todd and Alison South-Todd for first giving me the opportunity to study the archaeology of the Mediterranean islands, to the people of Kalavassos for their outstanding hospitality, and to those with whom I had the pleasure of sharing the experience, especially Ianthe Alexander, Claudia Chang and Chris Steele. I would also like to thank those colleagues working on islands in the Mediterranean and elsewhere, who have discussed their work with me over the years and given me much to think about, particularly Barbara Bender, Keri Brown, John Cherry, Steve Held, Caroline Malone, Manolis Melas, George Nash, John Schofield, Simon Stoddart, Stuart Swiny and Andrew Townsend.

I would like to thank the following for permission to reproduce the figures indicated: Ian Williamson (Figure 4.1), William Waldren (Figures 5.1 to 5.4), Ercole Contu (Figure 5.5), Ruth Whitehouse (Figure 5.6), Gary Webster (Figure 5.7), The Society of Antiquaries of London (Figures 5.9 to 5.11), Wolf-Dietrich Niemeier (Figures 5.16 and 6.7), Catherine Pèrles (Figure 6.1), Robert Tykot (Figure 6.2), Caroline Malone (Figures 6.3 to 6.6), E. & S. Sakellerakis (Figure 6.8).

Finally I would like to thank my parents, whose decision to bring me up on an island (Jersey) has left me with a lifelong fascination for the subject.

1

ISLANDS IN TIME:
AN INTRODUCTION

Islands have always held a particular fascination for people. For some, it is the isolation and relative security of island life that is attractive, the opportunity to escape from 'reality' as defined by other people and to reinvent it for oneself. For others, it is the challenge of coping in a harsh environment, with limited resources, that draws them towards islands, the opportunity to prove one's 'independence'. Whatever the attraction of islands, they feature prominently in European mythology (Circe's island in Homer's *Odyssey*, the 'Isles of the Blessed' of Celtic legend), literature (Shakespeare's *Tempest*, Defoe's *Robinson Crusoe*, Golding's *Lord of the Flies*) and popular culture (movies such as *Blue Lagoon*, musicals such as *South Pacific* and the radio series *Desert Island Discs*). Islands also feature prominently in archaeological literature: the islands of the Mediterranean in particular have attracted a great deal of research interest, from Sir Arthur Evans' (1921) and Sir Themistocles Zammit's (1930) early studies, to more recent work by John Evans (1971a), Renfrew (1972) and Renfrew and Wagstaff (1982). Turning to social anthropology, many of the classic ethnographic monographs concern island societies, from Malinowski's (1922) study of exchange systems in the Trobriand Islands, to Deacon's (1934) and Layard's (1942) studies of initiation rites and megalithic ritual in Malekula and Mead's (1943) study of sexuality and adolescence in Samoa. These studies, both archaeological and ethnographic, have had a profound influence on the theory and practice of contemporary archaeology. Until recently, however, little attention has been paid to the fact that these studies are specifically concerned with island societies. Today, archaeologists working in areas such as the Mediterranean and the Pacific are increasingly looking at the question of insularity. To what extent is it legitimate to use island studies as the basis for more general models of social structure and cultural change (Evans 1973)? Do island societies have specific features which set them apart from continental communities (Evans 1977; Kirch 1986)? What are the ecological and cultural effects of insularity (Fosberg 1963; Cherry 1981; Terrell 1986)? These are the questions which will be addressed in this book. It is, first and foremost, a book about islands rather than a book about

1

Mediterranean prehistory, but theoretical questions are best considered in relation to specific data, and the Mediterranean region has a large number of islands (Cherry (1981) lists 115) of varying sizes and degrees of remoteness (see Figures 1.1 and 1.2), many of which have relatively well understood archaeological sequences. The book will focus specifically on the archaeology of the Mediterranean islands between the end of the last Ice Age and the emergence of classical civilisation (arbitrarily set at 500 cal. BC). Because of its theoretical orientation, it is hoped that the book will be relevant to archaeologists and anthropologists outside the field of Mediterranean prehistory. At its core is the fundamental question of the relationship between human society and the natural environment: because insularity more so than any other environmental variable is clearly defineable, it is of particular interest in any attempt to understand this relationship. The first two chapters of the book will look at the theoretical background to island studies, and this will be followed by explorations of particular themes in Mediterranean insular prehistory: colonisation, ecology, elaboration and continuity, and networks of interaction. The examples used are necessarily and intentionally selective, the aim being to explore the relationship between insularity and culture, rather than to write a 'definitive' prehistory of the Mediterranean islands.

THE CONCEPT OF THE 'ISLAND LABORATORY'

John Evans (1973) must be credited with introducing discussion of insularity into the discourse of contemporary theoretical archaeology. He argued that islands could be seen as 'laboratories for the study of cultural process', since insularity 'tends to eliminate some of the variables which afflict the student of mainland groups, and whose effect is often so difficult to assess.'

Islands generally have a limited range of available resources. This allows the archaeologist to study ways in which a community has adapted to its natural environment, and can also provide unequivocal evidence for external contacts if resources not naturally present on a given island are found there in an archaeological context. In a mainland context, it is more difficult to define the 'environment' of a given community or to identify 'external' contacts, since this requires us to draw an artifical boundary around that community, and to define the area outside this boundary as 'external'. On an island, this boundary is not artificial, it is provided by the coast, and contact with other communities requires a deliberate sea voyage.

Evans stresses, however, that the sea, as well as dividing and isolating communities, may also be an effective means of communication between them, and suggests that this provides an opportunity 'to study the related development of small, discrete communities, to observe the mutual effect of their contacts, and to follow the development of differences between them'.

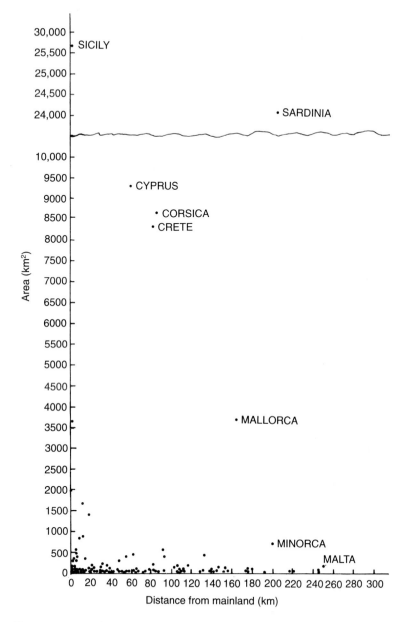

Figure 1.1 Distribution of the Mediterranean islands in terms of size (km²) and
distance from nearest mainland
Source: Cherry 1981

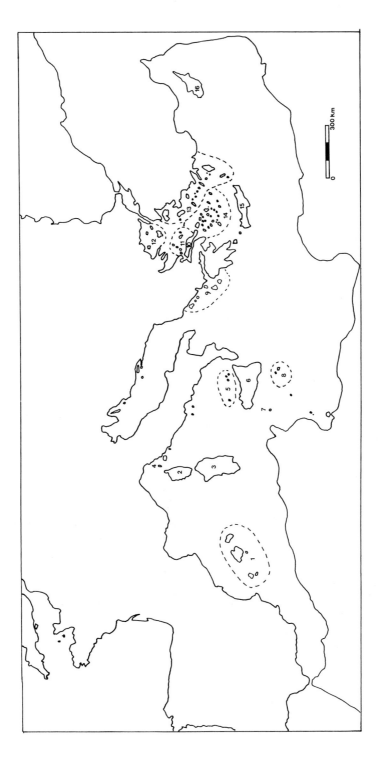

Figure 1.2 Map of the Mediterranean region, showing the major islands and island groups

Note: 1 The Balearic Islands; 2 Corsica; 3 Sardinia; 4 Elba; 5 The Aeolian Islands; 6 Sicily; 7 Pantelleria; 8 The Maltese Islands; 9 The Ionian Islands; 10 Euboiea; 11 The Sporadic Islands; 12 The Dodecanese Islands; 13 The North Aegean Islands; 14 The Cycladic Islands; 15 Crete; 16 Cyprus

In a later paper, Evans (1977) contrasts the Mediterranean situation with that in the Pacific, where islands are far more isolated, and where colonisation of an island was in many cases followed by development in total isolation, something which rarely, if ever, happened in the Mediterranean.

Evans' argument is primarily a methodological one: the cultural processes with which he is concerned are not specific to island communities, but affect all societies. Islands, however, provide an ideal opportunity to study these processes, since the relevant variables are fewer, and are more easily defined and controlled.

In developing the idea of the 'island laboratory', Evans follows in a tradition of research established by Charles Darwin. In *The Origin of Species*, Darwin (1968 [1859]) effectively used the Galapagos Islands as a laboratory of evolutionary processes. Darwin suggested that, following isolation, a plant or animal population would follow a separate evolutionary trajectory to its parent population on the mainland. He showed that, of the twenty-six species of land birds on the Galapagos, twenty-one of these were peculiar to the islands, and some were specific to particular islands in the Galapagos group: these species, although closely related to South American species, were none the less distinctly different. He saw this as conclusive evidence for independent evolution after isolation, and stated that this made no sense in relation to the 'creationist theory'.

The 'theory of island biogeography', developed by MacArthur and Wilson (1967) represents an attempt to quantify the variables involved in the colonisation of islands by animal and plant species, and in the subsequent evolution or extinction of these species. Like Darwin, MacArthur and Wilson were concerned with the possible role of islands as laboratories of evolutionary and ecological process. At the end of their book they suggest that an island, surrounded by water, can be seen as a model for ecosystems more generally: most ecosystems (forests, streams, heathland, etc.) are surrounded by areas of fundamentally different ecology, so that the colonisation of these ecosystems by animal and plant species, and the subsequent development of those species, will be affected by broadly the same variables as can be observed in an island context. In an island situation, however, these variables are more easily defined and controlled, making islands ideal laboratories for the study of ecological processes. Although MacArthur's and Wilson's theory was not developed with human communities in mind, it has had a considerable influence on archaeological studies of insularity (cf. Cherry 1981; Terrell 1986), and this is a theme to which we will return in Chapter 2.

FEATURES OF ISLAND SOCIETIES

The concept of the 'island laboratory' depends on the assumption that the processes affecting island communities are the same as those affecting mainland communities, or at least that the differences between them are differences of degree rather than of kind. There are, none the less, certain specific features of island communities which need to be examined.

A number of authors have pointed out that ecological limitation is a general feature of island life. Darwin (1968 [1859], 453) noted that: 'The number of species of all kinds which inhabit oceanic islands are few in number compared with those on equal continental areas.'

MacArthur and Wilson (1967) suggested that the biodiversity of an island will vary in direct proportion to a function of the island's size (i.e. larger islands can support a greater number of species) and in inverse proportion to a function of its distance from the mainland (i.e. more remote islands will tend to support fewer species). Reduced biodiversity in an island context is likely to require significant adaptation on the part of colonising human populations. Evans (1973) argues that this limitation makes islands ideal laboratories for the study of human adaptations to the natural environment, whilst Renfrew and Wagstaff (1982), in the introduction to their study of Melos, focus on this limitation in biodiversity as a 'significant characteristic of the island ecosystem'. For human communities, however, this limitation may potentially be offset by other factors. The reduced biodiversity of an island ecosystem applies only to terrestrial resources: the resources of the sea will be as rich as on any other coastal area, and may be equally important to human communities. A small island such as Malta or Melos allows all communities direct access to the sea, providing an important nutritional 'safety net', as well as an element of dietary diversity, which may actually give island communities an advantage over their land-locked counterparts. Islands may also have specific non-biological resources (such as obsidian on Melos) which may be used in exchange with communities on other islands and adjacent mainlands.

The relationship between a human population and its natural environment is a complex and reciprocal one. Having colonised an island, a human community must adapt to the new environment, otherwise, like any other animal community, it will face extinction. Human populations, however, may also have a dramatic effect on the environment itself. Evidence for significant resource depletion following human colonisation of an island has been noted, for example, on the Pacific island of Tikopia (Kirch and Yen 1982) and on the Reef Islands (Green 1976). The human impact on the environment may be apparent from its effect on the landscape itself as well as from declining biodiversity (Kirch and Yen *op. cit.*). In some cases the extent of this human impact may be so great as to provoke a collapse of the social system, or even threaten the survival of the human population, as argued by Bahn and Flenley (1992) in the case of Easter Island. In the Mediterranean, the degree of isolation of islands is very much less than in the Pacific, and ecological limitations consequently less extreme. The human impact on the environment, however, may none the less be significant: the most recent evidence from Cyprus (Simmons 1991) suggests that Mesolithic hunter–gatherer groups were responsible for the extinction of the pygmy elephant and hippopotamus on which their economy depended. There is no apparent continuity between the newly recognised Mesolithic of Cyprus and the earliest Neolithic of the island, and it seems very likely that hunter–gatherers either left the island, or followed their prey into extinction. The prehistoric human ecology of the Mediterranean islands will be considered in detail in Chapter 4.

If islands have certain specific ecological features, do human communities on islands also have particular cultural features? Evans (1973) notes that: 'Island communities often display a tendency towards the exaggerated development of some aspect of their culture, which is often connected with ceremonial.' This elaboration of culture in the island context, which Evans (1977) explores with reference to the Maltese temples, has also been noted and discussed by anthropologists working in the Pacific region (cf. Vayda and Rappaport 1963; Sahlins 1955). Easter Island is an obvious example of this phenomenon (Bahn and Flenley 1992), but we could also look, for example, at the megalithic traditions of Malekula (New Hebrides), as described by Deacon (1934) and Layard (1942). Whilst isolation can give rise to distinctive and original manifestations of culture, however, it can equally give rise to significant conservatism. Renfrew and Wagstaff (1982) argue that protection from outside competition in an insular context may give rise to: 'the preservation of archaic ... or possibly ill-adapted forms'. Melas (1985), for example, has convincingly shown that Minoan material culture forms survived on the island of Karpathos long after they had disappeared from the rest of the Aegean. These themes of cultural elaboration and conservatism will be explored in Chapter 5.

Much of the ethnographic literature on island societies has focused on elaborate networks of interaction and exchange, such as the 'Kula' ring of Melanesia (Malinowski 1922). Recent research (cf. Leach and Leach 1983) has shown how important these networks can be in relation to the social structure of the communities involved, and the political strategies of the individuals within them. In the prehistory of the Mediterranean there is extensive evidence for exchange between islands involving, for example, the exchange of obsidian from Mesolithic times onwards. Chapter 6 will look at the social and political significance of inter-island and island/mainland interaction in Mediterranean prehistory.

DEFINING INSULARITY

One of the first problems which must be addressed concerns the definition of insularity: what exactly do we mean by an island? A dictionary definition might describe an island as 'an area of land surrounded on all sides by water', but how meaningful is this definition in an archaeological context? With the benefit of modern maps and satellite images, we know that Australia is an island, but it is difficult to imagine pre-contact aborigines being aware of it as such. Euboia and Sicily are both islands, but are so close to the mainland as to render this insularity meaningless (the distances concerned would, for example, pose no problem to a reasonably proficient human swimmer). Certainly we would not expect to find any significant reduction of biodiversity on these 'islands', nor would we expect their prehistoric inhabitants to be in any sense isolated from their mainland neighbours. MacArthur and Wilson (1967) suggest that island size and distance from mainland are the most significant variables, at least in determining the extent of biodiversity on a given island, and these variables are also likely to be important in defining the extent to which the islands are perceived as such by their inhabitants. None of the Mediterranean islands are so large that they would not be seen as islands. Even on Sardinia, the largest truly 'insular' island, the maximum distance of any point from the sea is around 65 km. Equally, unlike the Pacific, none are so remote as to suggest that communities would be totally 'closed' (the most remote islands are the Maltese group, around 250 km from the nearest mainland). There is, none the less, a considerable diversity among the Mediterranean islands in terms of their size and distance from the mainland, as shown by Figure 1.1.

THE MEDITERRANEAN: A GEOGRAPHICAL AND ARCHAEOLOGICAL BACKGROUND

The Mediterranean sea extends from the Levant in the east to Iberia in the west (a distance of 3700 km), and from the Po estuary in the north to the Libyan coast in the south (a distance of 1800 km). Within this sea are 115 islands, most of which are relatively small (only Sicily, Sardinia, Cyprus, Corsica, Crete, Euboia and Mallorca have surface areas greater than 2000 km^2), and within sight either of an adjacent mainland or of another island which serves as a 'stepping stone' (the only exceptions to this being the islands of Lampedusa, Linosa and Lampione in the central Mediterranean).

The Mediterranean basin was formed by the separation of the African and European plates, around 150 million years ago. Geologically the area is dominated by limestones, though there are significant outcrops of igneous rock on Corsica, Sardinia, Sicily and Cyprus. Around much of the Mediterranean coast there is no extensive coastal plain, instead, mountains rise rapidly behind the coast – Mount Etna in Sicily, the Appenines of Italy, the Dinaric Alps in Dalmatia, the Pindus range in Greece, the Taurus Mountains of Anatolia and the Maritime Atlas of the Maghreb. The climate is relatively homogeneous, with hot, dry summers and mild winters. Vegetation is dominated by light woodland and (especially where this has been cleared by human activity), maquis. The soils are in many cases thin and, being calcareous, often lacking in groundwater, particularly in the dry summers.

The prehistoric sequence in the Mediterranean region has been admirably summarised by Trump (1980). Lower Palaeolithic material has been found on a number of Mediterranean sites, including Vallonet (de Lumley *et al.* 1963) and Terra Amata (de Lumley 1967) in France and Puig d'en Roca in Spain (Canal and Carbonell 1979). At Tautavel (de Lumley and de Lumley 1971) in France, and at Petralona (Stringer *et al.* 1979) in Greece, Lower Palaeolithic material was found in association with early hominid remains. This evidence shows that the Mediterranean region was first colonised by people around a million years ago. There is, however, no conclusive evidence for Lower Palaeolithic settlement on any of the Mediterranean islands, apart from Euboia (Micha-Sarantea 1980) and Sicily (Bianchini 1969) which, as we have already seen, cannot be considered as truly insular (and which, during cold periods of the Pleistocene, would have been joined to the mainland). A series of flint assemblages from the Anglona region of Sardinia have been claimed as Lower Palaeolithic (Arca *et al.* 1982), but these are unstratified assemblages, identified as 'Clactonian' purely on typological grounds. Middle Palaeolithic material has been found on the islet of Mikro Kokkinokastro (Cherry 1981), off the southwest of Halonissos (one of the Northern Sporadic Islands), and on the central Mediterranean island of Elba (Lanfranchi and Weiss 1973). Again, however, we must bear in mind the relevant palaeo-geographical factors: during glacial periods these islands would have been connected by land bridges to the continent, so that the presence of Palaeo-lithic material may not be relevant to the question of island colonisation. More recently, however, Middle Palaeolithic material has been found on Kephallenia (Kavvadias 1984) which, though joined to Ithaca and Zakynthos during much of the Pleistocene, was never part of the mainland.

Throughout the Upper Palaeolithic period, the climate of the Mediterranean area was significantly drier and cooler than it is today. Faunal evidence shows the presence of horse, reindeer, mammoth and marmot, whilst palynological evidence suggests a predominance of grassland, with small areas of pine and deciduous woodland (Trump 1980). Flint industries of broadly Aurignacian type have been identified in Italy, as at Grimaldi and Arene Candide, in Gibralter and at the site of Haua Fteah in Libya. At Grimaldi and Arene Candide, deliberate human burials were found, covered with red ochre and associated with shell beads. In the west Mediterranean the Upper Palaeolithic sequence closely follows the classic southwest French pattern, as illustrated by the cave of Parpallo in Spain, which has a stratified sequence from Gravettian to Solutrean and Magdalenian flint industries (Trump 1980). The Gravettian levels of this site are also of particular interest, owing to the presence of painted and engraved plaques. Upper Palaeolithic material is present on Sicily (Cherry 1981), at Levanzo in the Egadi Islands (Graziosi 1953), and perhaps also on Sardinia (Sondaar *et al.* 1984, but see also Cherry

1992). Of these, however, only Sardinia would have been an island at the time.

The Mesolithic period in the Mediterranean region is marked, as elsewhere in Europe, by the proliferation of microlithic flint industries, and by evidence for increasing economic specialisation. These trends are evident in the shell middens of the Capsian complex in the Maghreb, in the cave sites of the Castelnovian complex of southern France and in the Natufian (Final Mesolithic) of the Levant (modern Syria, Lebanon, Israel and Palestine). As far as the islands are concerned, there is evidence for Mesolithic occupation on Sardinia (Sondaar *et al.* 1984), Corfu (Sordinas 1970), Corsica (Lanfranchi and Weiss 1973, 1977), Mallorca (Waldren 1982; Kopper 1982) and Cyprus (Simmons 1991). All of these, with the exception of Corfu, must have been islands at the time of deposition. The evidence from the Franchthi Cave, on mainland Greece, also demonstrates the exploitation of obsidian from Melos from Early Mesolithic times (Perlès 1979). In general terms, therefore, the evidence for human activity on the Mediterranean islands prior to the Neolithic is limited, as Cherry (1981) suggests. The evidence from the Franchthi Cave demonstrates that Mesolithic groups in the Mediterranean were capable of undertaking sea voyages, but they seem not to have colonised the islands to any great extent. The significance of this will be further discussed in Chapter 3.

The earliest evidence for cereal cultivation and animal husbandry is from the Near East. The faunal assemblages from Natufian sites in the Levant suggest that hunter–gatherer communities were specialising increasingly in particular species: goat at Beidha and El Khiam, gazelle at Nahal Oren. The proliferation of querns in Natufian contexts also suggests increasingly intensive exploitation of wild cereals. By *c.* 8000 uncal. BC, settlements in the Levant were fully agricultural (though still aceramic): in some cases the new economy supported large stable communities, as at Jericho, where the walled settlement is estimated to have supported a population of around 2000. By 6250 uncal. BC (7200 cal. BC), the Neolithic way of life was firmly established in southeastern Turkey, as well as in the Levant, with important settlements at Hacilar, Can Hassan and Catal Huyuk (Mellaart 1975). The island of Cyprus, which had been occupied at a relatively early stage in the Mesolithic (Simmons 1991), was reoccupied during the eighth millennium cal. BC, the initial population having apparently either abandoned the island or become extinct. The Neolithic sites of Khirokitia (Le Brun 1984) and Kalavassos-Tenta (Todd 1987) are aceramic, and the early radiocarbon dates from Kalavassos-Tenta (7560 ± 130 BC; P-2972: 7040 ± 410 BC; P-2785) suggest the pre-pottery Neolithic groups of the Levant as a likely parent population.

Between 7000 and 5000 cal. BC, the Neolithic way of life became established over most of the Mediterranean area. The earliest Neolithic settlements in Thessaly and Macedonia have certain points in common with sites in

Anatolia: they are in many cases quite large settlements (as at Nea Nikomedea and Argissa Magoula), and the finds include painted pottery with clear Anatolian parallels. Whilst this has given rise to suggestions of a colonising population, there is other evidence to suggest a more gradual process of acculturation. At the Franchthi Cave (Jacobsen 1969; 1973), for example, there is considerable continuity between the lithic assemblages associated with the Mesolithic levels and those associated with the earliest Neolithic levels. It is probably wrong to attribute the spread of the Neolithic way of life into the Aegean area to a single process: colonisation events may well have occurred in some areas, whilst in other areas indigenous populations may have adopted all or part of the Neolithic 'package' as a result of contacts with farming societies. This suggestion is underlined by the possible existence of aceramic Neolithic horizons at sites such as Franchthi and Argissa Magoula. It is during this initial Neolithic period that we see the earliest conclusive evidence for human occupation on Crete (Cherry 1981; Broodbank and Strasser 1991).

In considering the adoption of the 'Neolithic package' in the central and western Mediterranean, Whittle (1985) contrasts the evidence from southern Italy and Sicily with that from the coastlands of central and northern Italy, southern France and eastern Spain. In southern Italy and Sicily the evidence is dominated by ditched enclosures, including large sites such as Passo di Corvo, Scaramella and La Quercia: the limited evidence available from these sites suggests a fully developed agricultural economy. These sites are also characterised by the presence of painted pottery, considered by some authors as evidence for a direct influence from the Aegean. Elsewhere, the evidence is dominated by the 'impressed ware' sites, which suggest a much more gradual adoption of the Neolithic way of life. At Coppa Nevigata, for example, despite the presence of pottery, the evidence suggests an essentially Mesolithic economy, based on the intensive exploitation of coastal resources. In southern France, at sites such as Châteauneuf-les-Martigues, there is clear evidence for continuity in the lithic assemblages from Mesolithic to Neolithic horizons. The evidence from southern French impressed ware sites also suggests that the herding of ovicaprids appeared at an earlier stage than cereal cultivation. Several of the west Mediterranean islands seem to have been colonised for the first time during the initial (impressed ware) Neolithic – most notably Lampedusa (Camps 1976) and the Tremiti Islands (Trump 1966). The earliest evidence of human occupation on Malta is at the site of Ghar Dalam (Evans 1971a), with a radiocarbon date of 4190 ± 160 bc (5260–4840 cal. BC). This period also saw the development of exchange networks within the central and west Mediterranean areas, based on obsidian (Hallam *et al.* 1976) from Sardinia, Pantelleria and Lipari. Sardinian obsidian is recorded from Early Neolithic contexts in Corsica (Phillips 1975), whilst both Pantellerian and Lipari obsidian are present in the earliest (Ghar Dalam) Neolithic contexts in Malta (Trump 1966). Pantellerian obsidian may also be present on Neolithic sites in Tunisia (Camps 1976).

By 5000 cal. BC, the Neolithic way of life was firmly established over most of the Mediterranean area, and a series of regional material culture traditions had been established: Halaf and Wadi Rabah in the Levant, Can Hasan in Southern Anatolia, Sotira in Cyprus (Mellaart 1975), Larissa and Dhimini in Greece, Scaloria, Ripoli and Fiorano in Italy, Chasseen in southern France and Almeria in Spain. The fifth and fourth millennia cal. BC saw important cultural developments in the Mediterranean area. First, in the Aegean, we see the appearance of large fortified settlements centred on a large open courtyard with a 'megaron' building, as at Dhimini and Sesklo (Theocaris 1973). Second, in the west Mediterranean, the period is marked by the appearance of megalithic monuments in Iberia (Leisner and Leisner 1943), and by the appearance of interrupted ditch enclosures in southern France (Phillips 1975). In Italy, the late fifth millennium cal. BC is marked by the appearance of rock-cut tombs, as at Serra d'Alto (Whitehouse 1972). In Malta, the earliest stone temples date to the Ggantija phase, between 3600 and 3000 cal. BC. All of these developments represent increasing investment of labour in communal projects, and could be taken as an indication of increasing social complexity and differentiation (cf. Whittle 1985). Many islands seem to have been colonised for the first time during the fifth and fourth millennia (Cherry 1981): the Ionian islands of Lefkas and Meganisi (Hope-Simpson and Dickinson 1979), the north Aegean islands of Lemnos and Thasos (Leekley and Noyes 1975), Chios and Samos in the eastern Aegean (Hope-Simpson and Dickinson 1979), Keos, Makronissos, Melos and Naxos (Hope-Simpson and Dickinson 1979) in the Cyclades.

The third millennium cal. BC marks the beginning of the Pharonic period in Egypt, and is also important in terms of the development of copper and bronze metallurgy in the Mediterranean region as a whole. Early copper and gold objects have been identified in Israel, as at Kfar Monash, in Cyprus (Early Cypriot I) and in the Aegean (Early Minoan I–II in Crete, Troy II in Western Anatolia, Early Helladic II in Mainland Greece). By the end of the third millennium, bronze was in widespread use throughout the eastern Mediterranean. Copper objects are also recorded in the Remedello and Rinaldone groups in Italy, the Fontbuisse group in southern France and the Bell Beaker complex of Iberia and the Balearic Islands. In terms of social and cultural change, the most dramatic developments are, perhaps, in the Aegean. In mainland Greece, for example, we see the appearance of large fortified sites, as at Lerna, Tiryns and Askitario: the so-called 'House of Tiles' at Lerna, and the circular building at Tiryns are large, and presumably public, buildings (Renfrew 1972). In Western Anatolia, the fortified settlement of Troy II is centred on a large 'megaron' – type building. In the Cyclades, similarly, we have the appearance of fortified settlements as, for example, at Kastri. The site of Phylakopi, though not fortified, is important as the earliest true town in the Cyclades (Renfrew 1972). Looking at the Aegean area as a whole during the third millennium, the evidence suggests an increasing

nucleation of settlement linked to increasing social differentiation: a process
which found its ultimate expression in the 'palace societies' of Minoan Crete
and Mycenean Greece. Although the most spectacular developments of this
period are in the Aegean, similar processes of centralisation and increasing
social differentiation can be identified in other parts of the Mediterranean
region. In Iberia, for example, the period is marked by the appearance of
large concentrations of megalithic monuments, as at Los Millares (Leisner
and Leisner 1943), where a group of eighty corbelled passage graves is associated with an important defended settlement. In Malta, the Tarxien Phase
(3000–2500 cal. BC) was marked by the construction of the largest and
most elaborate stone temples. These temples, however, went out of use in
the second half of the third millennium cal. BC. A significant number of
east Mediterranean islands were occupied for the first time during the third
millennium cal. BC: Cherry (1981) lists the Ionian Islands of Ithaca, Kythera
and Kephallenia, the Argo-Saronic Islands of Hydra, Poros, Salamis and
Spetsai, the east Aegean Islands of Lesbos and Nisyros and the Cycladic
Islands of Delos, Despotiko, Donoussa, Heraklia, Ios, Keros, Kouphonissia,
Rheneia, Schinoussa, Siphnos, Syros and Tinos.

The second millennium cal. BC was marked by the appearance of 'palace
societies' in the Aegean. The earliest palaces of Crete appeared at around
2200 cal. BC. These palaces, at Knossos, Phaistos, Mallia and Kato Zakros,
are integrated complexes of buildings with monumental elements and large-
scale storage facilities. It is in this context that writing appears for the first
time in Europe: initially 'hieroglyphic' scripts and later Linear A, found on
clay seals and tablets. Throughout the Aegean Bronze Age, writing seems to
have been used essentially for administrative, rather than for literary purposes.
During the 'New Palace period' (1700–1450 cal. BC), there were five palaces
in Crete (Knossos, Phaistos, Mallia, Kato Zakros and Khania), each controlling a territory of 1000 to 1500 km². Most of these palaces had associated
urban settlements. In the final period (1450–1370 cal. BC) all the palaces
were destroyed, with the exception of Knossos, which seems to have established a position of political hegemony. Knossos itself, however, was destroyed
at c. 1370 cal. BC, and with it the Minoan palace civilisation. The palaces
of mainland Greece, in contrast to those of Crete, are incorporated within
defended citadels. Regional centres, many of them fortified, had already developed during the third millennium, as at Lerna, Tiryns and Askitario. From
around 1650 cal. BC, however, the citadel of Mycenae became dominant.
The shaft grave cemeteries associated with the early citadel show a far greater
concentration of wealth than has been identified on any of the earlier or
contemporary citadels. The true palace societies of mainland Greece, however,
developed following the collapse of their counterparts in Crete. The defensive walls of citadels such as Tiryns, Argos and Mycenae enclose the administrative, ceremonial and residential centres of these Bronze Age communities.
As with the earlier Cretan palaces, they include storage facilities (e.g. the

'Wine Magazine' at Pylos) and evidence for an administrative function (e.g. the 'Archive Room' at Pylos, with large numbers of Linear B tablets). The citadels of Mycenean Greece were abandoned between 1200 and 1050 cal. BC, in the context of a more general disintegration of Mycenean cultural unity. Whilst previous generations of archaeologists saw this as evidence for an invasion from the north, it is now clear that there was no abrupt, violent and single end to Mycenean culture. All of the citadels were destroyed, but at different times, and sometimes more than once (at Mycenae itself there are at least three destruction phases, in LHIIIb1, LHIIIb2 and LHIIIc). On balance the evidence suggests a protracted period of unstable conditions, probably caused by internal social factors rather than by invasion. The economic and historical basis of Bronze Age civilisation in the Aegean has been extensively debated. Early interpretations focused on the evidence for international trade, particularly in Crete, giving rise to the notion of a Minoan thalassocracy, an elite whose wealth and status depended almost entirely on control of the seaways on the eastern and central Mediterranean. This view of Aegean civilisation has been criticised, for example, by Renfrew (1972), who argues that the palace civilisations of the second millennium cal. BC emerged in the context of processes which can be traced back into the third millennium. International trade develops at a relatively late stage in this process and cannot, therefore, be seen as the prime mover. In Renfrew's view, the Bronze Age civilisations of the Aegean were based on the control of local resources, specifically wheat, olives and grapes (the principal elements of 'Mediterranean polyculture', which developed in the Aegean region at the beginning of the third millennium cal. BC). Outside of the Aegean region, in Cyprus the second millennium cal. BC was marked by an explosion of international trade, based on the export of copper. From *c.* 1600 cal. BC we see the appearance of imports from Syria, Egypt, Crete and the Levant in Cypriot tombs and, linked to this explosion of trade, the appearance of towns, as at Kalopsidha and Nitovikla (Trump 1980). Mycenean pottery appeared in Cyprus at around 1450 cal. BC, and the island appears to have been settled by Mycenean refugees following the collapse of Aegean civilisation at the end of the second millennium. The city of Enkomi, for example, was violently destroyed before being refounded as a defended Mycenean city. In the central and western Mediterranean, the influence of the Aegean civilisations is now generally recognised to have been far less important than was once thought. Although significant quantities of Mycenean pottery have been recovered from Scoglio del Tonno, in southern Italy, and characteristic 'oxhide' copper ingots have been found in Sardinia, there is no convincing evidence for an Aegean presence further west than this, and Blance's (1961) idea of Aegean colonists in Iberia can no longer be sustained (Renfrew 1967). In Sardinia, the second millennium is marked by the development of 'nuraghic' architecture. Nuraghi are stone buildings incorporating large round towers, generally considered (cf. Trump 1980) to be defensive in function.

The limited dating evidence available suggests that the earliest nuraghi date to between 2000 and 1500 cal. BC. Webster (1991) argues that these structures (there are around 7000 in total) were the fortified residences of local petty chiefs. Although Webster (*op. cit.*) has argued convincingly that the construction of these buildings would not require the existence of a highly stratified feudal-type society (cf. Lilliu 1959a,b), it is none the less reasonable to suggest that the appearance of these sites represents a significant increase in social differentiation. Associated with the nuraghi are a series of megalithic gallery graves known as 'tombi di giganti', characterised by the presence of a crescentic forecourt and, in many cases, a carved and panelled portal stone. In Corsica the period is characterised by the appearance of 'torri', stone buildings similar to the Sardinian nuraghi. In the southern coastal area of Spain, the second millennium cal. BC saw the development of the 'Argaric culture', characterised by large, fortified settlements, as at El Argar and El Oficio. Burials associated with these settlements suggest a significant degree of social differentiation. The Balearic Islands of Ibiza, Menorca and Formentera seem to have been settled for the first time either at the end of the third or at the beginning of the second millennium cal. BC (Topp *et al.* 1979). In the second half of the second millennium cal. BC, early rock-cut tombs in the Menorca were replaced by 'naveta' monuments – boat-shaped tombs of drystone construction. One such monument, at Es Tudons, has a two-storeyed chamber which contained large quantities of human remains. The 'talayot' monuments of the Balearic Islands, which appear to be slightly later than the naveta, are drystone buildings with massive walls, in some respects similar to the Sardinian nuraghi and the Corsican torri. Associated with the talayotic complexes are the 'taula' monuments, consisting of a single horizontal stone slab resting on a single upright. The function of these unique monuments is not known, though they are generally assumed to have had some ritual or ceremonial significance. In terms of island colonisation, the most significant activity in the second millennium cal. BC was in the Balearic Islands, as discussed above. Cherry (1981) also mentions, however, the Ionian Island of Zakynthos, the Dodecanese Islands of Lipsoi and Patmos, and the Cycladic Islands of Therassia and Anaphi, where archaeological evidence suggests initial colonisation during this period.

With the collapse of Aegean civilisation in the late second millennium cal. BC, Greece became isolated from developments further east. Iron was introduced to the Aegean for the first time at the end of the second millennium, probably from Anatolia, but the 'sub-Mycenean' and later geometric pottery styles show little evidence of contact with the eastern Mediterranean, and it seems clear that the international trade networks of the second millennium collapsed along with the Aegean civilisations of which they were part. Similarly, there is little evidence for any continued Aegean influence in southern Italy, or in the central Mediterranean more generally. Between 1000 and 750 cal. BC, communities in the central Mediterranean followed their own cultural trajectory, though with increasing Phoenician influence, following the establishment (according to historical sources) of colonies in North Africa (Utica) and Spain (Cadiz) at around 1100 cal. BC. On the Italian mainland, the first two centuries of the first millennium are characterised by rock-cut tombs and cremation cemeteries, and by bronze hoards of entirely indigenous character. In Sardinia, this period saw the elaboration of the nuraghic complex, with the appearance of larger fortified centres with multiple towers, as at Palmavera and Barumini. The period is also marked by the appearance of votive depositions of bronze figurines associated with sacred wells or springs, as at Sant' Anastasia and Santa Vittoria di Serri (Trump 1980). The earliest archaeological evidence for a Phoenician presence in the Central Mediterranean is an inscription from Nora in Sardinia, with a ninth-century inscription (Trump 1980). In the Balearic Islands, the construction of talayots and taulas continued into the first millennium, whilst in the Iberian mainland the period is marked (at least in northern Spain) by the appearance of urnfields. Whilst the Iberian urnfields clearly belong to a cultural complex which extends from central Europe, through northern Italy to Iberia, it is questionable whether they relate to the 'Celtic invasions' alluded to by Herodotus and suggested by some archaeologists (cf. Savory 1968).

From the eighth century onwards, the archaeological and art-historical evidence from the Aegean suggests a re-opening of trade contacts with the eastern Mediterranean. The 'orientalising phase' in archaic Greek art is marked by the appearance of animal depictions (including mythical animals such as the sphinx and griffin) and floral motifs (lotus and palmette) derived from Anatolian, Near-Eastern and Egyptian traditions. This period also sees a phase of urbanisation and centralisation, culminating in the development of the Greek city-state. By the mid-eighth century, contacts between the Aegean and the central Mediterranean had also been re-opened, with the establishment of Greek colonies at Pithecusae, Cumae, Syracuse and elsewhere in southern Italy. In the seventh and sixth centuries, this Greek influence was extended to the western Mediterranean, with the appearance of trading colonies at Massalia (Marseilles), in southern France and at Emporion (Ampurias) in northern Spain. The same period is marked by the appearance of Phoenician colonies at Motya in western Sicily, and at Sulcis in Sardinia. The appearance of Greek and Phoenician colonies in the central and western Mediterranean had a dramatic impact on the native societies of these areas. Trade was one of the principal motives for the establishment of these colonies, and the control of trade between the colonies and their hinterland became an important part of the social strategies of local elites. In addition to controlling trade, these local elites sought to enhance their status by emulating aspects of Greek and Phoenician culture, including pottery styles, metalwork and architecture. The effect of this interaction is most clearly seen in Italy, where increased competition is manifested in the development of the Etruscan city-state, with urban centres at Tarquinia,

Cerveteri, Veii, Vulci and Vetulonia. Similarly, in Iberia and southern France, we see the appearance of native urban centres (Trump 1980), as at Ullastret (Spain), with its circuit of defensive towers and Ensérune (France). By this stage, however, we have reached the end of prehistory and, consequently, the end of this study. The approaches considered in this book are, in most cases, probably not appropriate to the market societies of the Classical world and, even if they were appropriate, these societies lie outside the author's field of competence. We will therefore confine ourselves to a consideration of Mediterranean island prehistory, from the end of the last Ice Age up to the emergence of Classical civilisation.

A THEORETICAL FRAMEWORK

The past two decades have seen the collapse of the monolithic culture–historical approach (Renfrew 1973) which dominated archaeological thinking in the first half of the twentieth century, and the emergence of increasingly polarised 'processual' and 'post-processual' approaches. The differences between these two approaches are both methodological and theoretical.

On a methodological level, the processual approach (cf. Renfrew 1972) is characterised by an emphasis on objectivity and theory-testing. Hypotheses and explanations are considered to be useful only if they can be subjected to some form of independent 'testing' against the data. Whilst some early processual models were framed in the context of a strict Hempelian positivism, a 'falsificationist' model derived from Popper (1959) is accepted by most archaeologists working within the processual tradition. Archaeologists working within the post-processual tradition have, in general, been far less concerned with methodology and most have, implicitly or explicitly, followed a 'realist' epistemology (cf. Wylie 1982), in which archaeology is seen as a necessarily interpretative (and therefore subjective) discipline.

To a large extent, the theoretical differences between the processual and post-processual schools are pre-determined by the differences in methodology. Processual models tend, therefore, to be materialist in orientation (since we are dealing with material data) and, in some cases, functionalist (since it is easier to test a hypothesis concerning 'function' than one concerning 'meaning'). The development of the post-processual school in the 1980s was, in large part, a reaction against the perceived materialism and functionalism of the 'New Archaeology'. Archaeologists working within the post-processual tradition, therefore, have tended to favour idealist interpretations rather than materialist ones, and have focused on cultural 'meaning' rather than 'function'. Processual archaeology is, by definition, diachronic, focusing on the explanation of cultural dynamics (cf. Renfrew 1972). Post-processual archaeology, being influenced by structuralism, has in many cases concentrated on 'statics' rather than 'dynamics', on 'synchrony' as opposed to 'diachrony' (cf. Scholte 1979). Archaeologists have always borrowed theory from other disciplines and, whilst processual archaeology has tended to borrow from the natural sciences and from geography, post-processual archaeology has been more inclined to borrow from sociology and post-modernist philosophy.

These opposing theoretical traditions have influenced island studies as much as other areas of archaeological research. The whole concept of the 'island laboratory', as developed by Evans (1973) depends upon a hypothetico-deductive methodology, and the 'theory of island biogeography', borrowed from evolutionary biology (MacArthur and Wilson 1967) by archaeologists such as Cherry (1981) and Terrell (1986) is typical of the processual approach. For their part, post-processual archaeologists have generally avoided discussion of questions such as insularity, and its effects on human communities, considering this to be part of the materialist and determinist agenda of their processual rivals. This brings us to one of the central problems of the processualism vs. post-processualism debate: one of the reasons that archaeologists working within these different traditions come to fundamentally different conclusions is that they are asking fundamentally different questions. It is, of course, the prerogative of the individual researcher to decide on the questions which she/he wishes to pose, but surely we need an approach to archaeology which allows us to answer any question we may choose to pose, and which has the flexibility to recognise that different methodologies may be appropriate to different questions. The methodological straitjacket of the traditional 'processual' archaeology denies almost any possibility of answering questions about cultural meanings, belief systems, and the communicative dimensions of material culture, whilst the ideological straitjacket of 'post-processual' archaeology will not allow consideration of environment, ecology and the material basis of culture. Given this impasse, it is not surprising that some recent work (cf. Bradley 1990; Patton 1993) has tried (at least implicitly) to move beyond the processual/post-processual dichotomy in integrating the types of question and the types of approach usually considered as typical of these two different traditions. This, also, is the aim of this book. The approach taken has certain features in common with a 'processual' approach (an emphasis on social dynamics, a concern with the relationships between environment, culture and society, and with the material basis of social relations and social change), and other features in common with a 'post-processual' approach (a realist epistemology, a concern with human agency and intentionality, and with the ways in which material culture is actively used in the reproduction and transformation of social relations). It is in the context of this synthesis of approaches that the 'island sociogeography' model (see Chapter 2) has been developed.

2

APPROACHES TO ISLAND ARCHAEOLOGY

Charles Darwin (1969 [1859]), in looking at the flora and fauna of the Pacific Islands, identified certain recurrent features of island ecosystems. First, he recognised that the number of species present on islands was relatively small, and that this was particularly true of remote islands:

> The number of species of all kinds which inhabit oceanic islands are few in number compared with those on equal continental areas ... New Zealand, for example ... contains altogether only 960 kinds of flowering plant ... We have evidence that the barren island of Accension aboriginally possessed less than half a dozen flowering plants.
>
> Darwin 1968 [1859], 453–454

Second, he noted that: 'Although ... the species are few in number, the proportion of endemic kinds (i.e. those found nowhere else in the world) is often extremely large' (*Ibid.* 454). He noted, however, that these species, although unique to their particular islands, were closely related to those on other islands and adjacent mainlands, suggesting divergent evolution from a common ancestor.

Finally, he noted that certain types of animal (land mammals, frogs, toads, etc.) were completely absent from true oceanic islands, despite the fact that these islands could, in fact, support such species, as shown by the survival of introduced mammals.

Darwin was not interested in island populations for their own sake: for him, the islands of the Pacific offered the opportunity to test (and ultimately falsify) the prevailing creationist paradigm of the time. If, as the Old Testament suggested, God had created the world, and all species within it, in seven days, we would not expect to see evidence for divergent evolution of island species from a common mainland ancestor. Nor would we necessarily expect to see certain types of animals absent from islands which could in theory support them.

THE THEORY OF ISLAND BIOGEOGRAPHY

The theory of island biogeography of MacArthur and Wilson (1967) represents an attempt to define and quantify the factors involved in the colonisation of islands by animal and plant species, the survival or extinction of those species and their subsequent evolutionary development. Their approach was founded partly on Neo-Darwinian evolutionary biology, and partly on the more recent science of ecology. Like Darwin, they were not concerned with islands for their own sake, but with the possibilities islands offered as 'laboratories' for the study of more general evolutionary and ecological processes. Specifically, they argue that islands provide a good model for ecosystems more generally, since most ecosystems (woods, streams, dunes, marshes, etc.), which support a particular range of species, exist as 'islands', surrounded by very different ecosystems, potentially as hostile to those species as the sea is to terrestrial plants and animals.

According to MacArthur and Wilson (*op cit.*), islands have three specific qualities:

1 They are relatively small.
2 The sea surrounding them acts as a barrier to species dispersal.
3 They may have altered climatic variability as compared with adjacent mainlands.

These characteristics of islands have a number of effects on animal and plant species, notably reduced habitat variety, reduced immigration and higher extinction rate, differential dispersal mechanisms and altered population stability.

The area and distance effect

Central to MacArthur's and Wilson's theory are the concepts of area and distance effect. The probability of initial colonisation of an island by an animal or plant species varies in direct proportion to some function of the size of the island, and in inverse proportion to some function of its distance from the mainland (1967, Chapter 2). Given that island colonisation by plant and animal species is generally accidental, a species is more likely to arrive on a large island close to the mainland (such as, for example, Sicily) than on a small and remote island (such as Pantelleria). The size of the island also affects the number of species which can be supported. Thus, although Elba is closer to the mainland than Sardinia (Figure 1.2), we might expect Sardinia, being very much larger, to support a greater range of species. The existence of 'stepping-stone islands' between the mainland and a given island may assist species dispersal. Thus, whilst Crete lies at a considerable distance from either the Greek or the Anatolian mainland (Figure 1.2), the existence of intermediate islands (Kythira and Andikythera between Crete and Greece, Kos, Karpathos and Rhodes between Crete and Anatolia) means that the greatest extent of water that would have to be crossed in one journey would be between 40 and 50 kilometres. If the stepping-stone islands are small, however, they may also act, through competitive exclusion, to filter species from the migrating pool. Andikithera is a relatively small island, which can support only a limited number of species, so that a new species arriving from the Greek mainland via Kythera might fail to become established there and would consequently be unable to move on to Crete, an island large enough to support a wider range of species.

Species equilibrium

MacArthur and Wilson (*Ibid.* Chapters 3 and 4) go on to argue that, as a function of the 'area effect', the number of species present on an island will generally represent an equilibrium between the arrival of new species and the extinction of those already there. Thus species may come and go, but the number of species remains approximately the same. An island is considered to be 'closed' to a given species if it is excluded by competitors already in residence, or if its population size is held so low that extinction occurs much more frequently than immigration. When a species invades a new island it generally meets a new environment. It adapts either by contracting its niche (if there are more competitors), or by expanding it (if there are less).

The founder effect

The idea of the 'founder effect' (Mayr 1954) is that, since a colonising population will normally consist of only a few individuals, they will carry with them only a small part of the gene pool extant within the parent population. This may lead to rapid genetic divergence between island and mainland populations (the phenomenon noted by Darwin).

Archaeological applications of the biogeography model

Although MacArthur's and Wilson's model was not, in itself, concerned with human island populations, it has been extensively used by prehistorians in the recent literature, in attempting to explain patterns of island colonisation and the evolution of island communities. As Cherry (1981) points out, humans are animals, and are subject to many of the same ecological and evolutionary processes that affect other species. Island biogeography theory has been employed in archaeological studies of both the Mediterranean (Cherry 1981) and the Pacific (Kaplan 1976; Terrell 1977, 1986).

Archaeological applications of biogeography theory have in many cases focused on the 'area and distance effects'. Cherry (1981) looks at the chronological evidence for the initial colonisation of the Mediterranean islands, and shows a close correlation between the pattern of island colonisation observed in the archaeological record and that expected on the basis of the MacArthur and Wilson model: large islands, close to the mainland, were settled first (Corfu, Corsica and Cyprus in the Mesolithic, Crete in the Early Neolithic), and smaller, more remote islands at a later stage (The Maltese Islands and Pantelleria in the late sixth millennium cal. BC, many of the smaller Aegean Islands in the fifth and fourth millennia).

Kaplan (1976) and Terrell (1977; 1986) have both argued that the distance and area effects can be used, not only to model the processes of island colonisation by people, but also the extent of interaction between island communities. This takes the biogeography model somewhat beyond MacArthur's and Wilson's original concept since, unlike humans, animal and plant species, having colonised an island, generally develop in complete isolation from populations on the mainland and other islands. Both Kaplan and Terrell use the mathematical technique of proximal-point analysis. Terrell (1977) uses first, second and third proximal-point analysis to model patterns of contact in the Solomon Islands. Single points are substituted for small islands, and the largest islands (which are elongated) are reduced to a straight line, defined by three points (a mid-point and two end points). Having determined these points, three lines are drawn from each point to the three points closest to it, expressing the working hypothesis that these would be the preferred networks of communication (Figure 2.1). Terrell (1986) also cites the 'gravity model' of human interaction, which suggests that an attracting force is created by the population masses of two areas, and that friction against interaction is created by intervening space. Thus, interaction varies directly with some function of population size, and inversely with some function of distance. This is important since it suggests that, in modelling interaction (as opposed to colonisation), population size should replace island area as one of the two main variables. In an archaeological context, however, population size is very difficult to estimate, and area has generally been used, on the assumption that these two variables are related to one another.

The concept of species equilibrium may also be relevant to studies of human island populations. Humans are, in general, more efficient than most other species in eliminating competition: this explains their success in colonising all five continents in a relatively short (in evolutionary terms) period of time. Furthermore, since the beginning of the Neolithic, humans have not travelled alone, but have brought with them a whole range of other species: domesticated cattle, sheep, pigs, dogs, rabbits and fowl, cultivated cereals, yams and other plants, as well as unwelcome stowaways such as rats. We would therefore expect the human colonisation of an island to be followed, either by the extinction of that population itself (as seems to have happened on the Galapagos, which were uninhabited at the time of European contact, but which have produced archaeological evidence for prehistoric activity (Heyerdahl and Skolsvald 1956), or by the extinction of a number of other species. There are numerous examples of the latter incidence, of

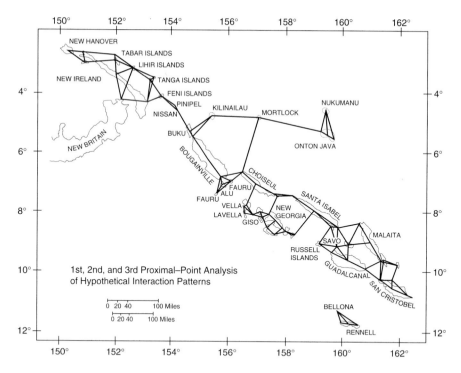

Figure 2.1 A proximal-point map of the Solomon Islands and New Ireland

Source: Terrell 1986

which perhaps the most famous is the extinction of the Dodo, following the European colonisation of Mauritius. Kirch and Yen (1982) have outlined archaeological evidence for a significant decline in both the numbers and diversity of shellfish, birds and fish (including the extinction of a flightless bird, the Megapode) following the initial human colonisation of the Polynesian outlier of Tikopia, and Green (1976) has used studies of shell-fish remains to show a considerable reduction in species diversity following the human colonisation of the Reef Islands.

It could also be suggested that the 'founder effect' is in some respects relevant to archaeological studies. Certainly it will be relevant in studying the genetics of human island populations (though its effect may be negligible if the island population maintains links with mainland and other island populations through intermarriage). Vayda and Rappaport (1963) employ an argument similar to that advanced by Mayr (1954), suggesting that a founder population is likely to carry with it only a small proportion of the *cultural* characteristics present within the parent population, and that these characteristics may

23

become exaggerated and accentuated through isolation, giving rise, for example, to elaborate monumental ritual in remote islands such as Easter Island.

LIMITATIONS OF THE ISLAND BIOGEOGRAPHY MODEL

Models derived from island biogeography theory have had considerable influence in recent studies of island archaeology, and have been extremely useful in focusing attention on the effects of insularity on the development of prehistoric communities. Such models, however, do have their limitations: MacArthur's and Wilson's (1967) model has been criticised by biologists (cf. Berry 1979; Williamson 1981) on the grounds that it ignores important aspects of the biology of individual species. These criticisms may be especially pertinent in considering the use of the theory to explain the development of human island communities. Although, in many respects, the factors affecting human island populations are essentially similar to those affecting other species, in other respects they are quite different.

There are a number of differences between island populations of plants and animals on the one hand, and people on the other. First, the colonisation process itself is in many cases quite different: colonisation by plants and animals is generally accidental, whereas colonisation by humans may be either accidental or (more often) deliberate. Deliberate exploration and colonisation of islands by people involves a range of variables which do not affect other species (other than those accidentally or deliberately brought to an island by people):

- Possible social and political motives for exploration and colonisation.
- The demand for particular resources which may be present in an island context (e.g. obsidian on the Mediterranean islands of Melos, Lipari and Pantelleria).
- The level of maritime technology and the type of sea-craft available.
- The extent of knowledge of tides, currents and other navigational factors.

Second, there is an ecological distinction between human and other animal populations. One of the fundamental tenets of the MacArthur and Wilson (1967) model is that the biodiversity of a small and remote island will be less than that of a larger island or continental area, and that colonising species need to adapt to this reduced biodiversity. This applies, however, only to terrestrial resources (on which most terrestrial species depend). The resources of the sea, which may be equally important to human groups, will be no less diverse and no less abundant than elsewhere, and the area of coastline available to each community will in many cases be greater. The exploitation of coastal resources, therefore, may allow human colonists to be significantly more successful than the island biogeography model would suggest.

Third, whilst an island population of an animal or plant species, once established, develops in complete isolation from its parent population, human populations may have boats, and may continue to have links with communities on other islands, involving both genetic and cultural exchange. This is particularly the case in a situation such as that in the Mediterranean region, where the vast majority of islands are relatively close to the mainland, and may lead to a very different pattern of insular development than would be expected in a totally isolated population.

One of the fundamental differences between humans and other animals is the proportion of learned as opposed to instinctive behaviour patterns. The MacArthur and Wilson model is essentially concerned with the transmission, survival and evolution of inherited characteristics in insular contexts, whereas the archaeologist or anthropologist must also be concerned with the transmission, survival and evolution of cultural characteristics. The processes by which these characteristics are reproduced and transformed are fundamentally different from those affecting genetic traits. This has implications not only in considerations of the 'founder effect', but in all areas of island life. Much of MacArthur's and Wilson's (1967) book is given over, for example, to discussion of the demography of island populations. What are the chances that a founder population of a given size will be able to successfully reproduce itself? And, having established an island population, what will be the ecological effect of population expansion? Whilst the demography of an animal or plant population can be established using a relatively simple mathematical model, the demography of a human population will be complicated by a range of cultural factors, notably social institutions such as marriage and incest rules, polygamy, infanticide and birth control. Human cultural behaviour also involves the establishment, reproduction and transformation of complex (and often asymmetric) relationships between individuals, which are articulated through the manipulation of the material world. The negotiation and articulation of such relationships may provide a motivation for exploration and colonisation (see Chapter 3), and may also have a marked effect on the relationship of a human community to its environment. Human populations frequently establish social relations through the production and (unequal) redistribution of a surplus, which may be used to finance competitive feasting, to exchange with other communities or for monument construction. The demand made on the environment by an animal or plant species varies in direct proportion to its population size, whereas human pressure on the environment may be as much a function of these social demands as of demographic expansion.

In conclusion, therefore, whilst the theory of island biogeography has made a considerable contribution to archaeological studies of island populations, it cannot account for the full complexity of the factors involved in the human colonisation of islands, and in the subsequent development of island societies. To explore, in more detail, the effect of insularity on human populations we need to develop a theory of island sociogeography, focusing specifically on the cultural dimensions of island life. The need to move beyond the island biogeography model has, in fact, been widely recognised in the recent literature, and many authors (cf. Kirch 1984, 1986; Kirch and Yen 1982; Stevenson 1986) have gone much further in their conceptualisation of island social formations and their development over time. In most cases, however, these authors have been concerned with specific case studies, and have not attempted to develop a body of general theory to explain the relationships between insularity and social structure. To a large extent, this may reflect a reaction against the ecological determinism implicit in the island biogeography model itself, and an unwillingness to replace it with an equally deterministic model. A general body of theory to explain the effects of insularity on the development of human societies would necessarily be different from from one developed in evolutionary biology and non-human ecology, and would have to avoid the determinism which has characterised some recent work in the field (cf. Terrell 1986). In fairness to the advocates of island biogeography theory in archaeology, it should be stressed that they have in many cases recognised the limitations of the model. Cherry (1990), discussing his earlier work, makes it clear that he suggested the model 'merely as a useful exploratory strategy' and describes his (1981) paper as 'a deliberately provocative use of biological determinism', stressing that 'a more sophisticated and balanced view is obviously necessary'. It is in an attempt to explore the implications of such a view that this book has been written.

OUTLINE OF A THEORY OF ISLAND SOCIOGEOGRAPHY

One of the central questions of archaeology concerns the nature of the dynamic relationships between human communities and their environment. Certainly this question was historically central to the development of the 'new archaeology' in the 1960s (cf. Binford 1962, 1965). The agenda of the 'new archaeology' has been systematically criticised by advocates of the 'post-processual' approach, largely because of the deterministic way in which these relationships were modelled, and because of the essentially passive role which these models accorded to human agency. Cultural systems were seen as an 'extra-somatic means of adaptation' to the environment. For some, the functionalism and determinism inherent in this approach have discredited the entire project of understanding the relationships between human communities and their environment. The challenge, however, is to develop a new way of understanding these relationships, within the framework of an approach which avoids functionalism and determinism, and which recognises the active role of human agency. Such an approach would look at the way in which the environment is manipulated and transformed through social practice, and at the ways in which specific aspects of a given environment are used in the articulation and transformation of social relations. Whilst there are many environmental variables that would be relevant to such a consideration, insularity is perhaps the easiest to define and quantify.

A study of insularity, and its effect on human societies, would necessarily begin with a consideration of the human colonisation of islands. Why did people colonise islands in the first place? Population pressure is an obvious possibility: if the population of a given community grows beyond its carrying capacity, one solution is for a group of younger people to leave, in the hope of establishing a new community, either in another mainland area or (particularly in the Mediterranean) on an island. Firth (1961) records an example of this in the Pacific, in which a dispute over land on the island of Tikopia resulted in the expulsion of one lineage (the Nga Faea) by another (the Nga Ariki). The Nga Faea left in search of a new land, with the women and children in canoes and many of the men swimming alongside. Although there is no record as to whether they succeeded in finding a new home, their chances would have been slim: the seas around Tikopia are infested with sharks, the distances between islands in that part of the Pacific are considerable and, in any case, by the time of this event there was little unclaimed land to be found anywhere in Oceania. The chances of survival of a comparable prehistoric group in the Mediterranean, however, would have been very much greater. There may also have been social and political reasons for island colonisation. Kirch (1984) suggests that internal social competition may have been a factor: the defeated rival of an established ruler, or the younger brother of an hereditary chief, may have preferred to take the risk involved in setting sail to find a new island, where he could establish his own authority, rather than accepting a permanent state of subjugation. In a maritime society, exploration and colonisation may provide young people with the opportunity to bypass the normal power structures within their own community. This may also serve the political interests of the ruling group within the parent community, since potentially dangerous rivals can be removed to a different arena. We might therefore expect to find a correlation between the frequency of island colonisation in a given area, and the demography and social formation of the parent community. As Cherry (1981) has shown, many of the Mediterranean islands were not colonised until a relatively late stage, and population pressure and political ambition are both likely to have been factors in their colonisation. Certainly these factors were significant in the later establishment of Greek colonies in the central and western Mediterranean, though in this case the territories which were colonised were, in most cases, already occupied. The colonisation of the Mediterranean islands will be considered in detail in Chapter 3 of this book.

Much of the recent literature on island societies has focused on the ecological dimensions of island settlement, and this will be discussed in Chapter 4. MacArthur and Wilson (1967) stress that island colonisation requires an adaptation to a new environment, generally characterised by reduced biodiversity and, according to Bates (1963): 'One cannot help but suspect that small insular populations are . . . more aware of resource needs, or have developed necessarily good cultural adaptations to resource needs'.

Demography is one of the most significant factors in a human population's relationship to its environment. MacArthur *et al.* (1976) have carried out a computer simulation to show the probability of survival and reproduction of a founder population. As one would expect, the size of the initial founder population is a major factor: with an initial population of six (assumed to be three couples of reproductive age) the probability of extinction was calculated at 77 per cent, compared with 19 per cent for an initial population of seven couples. The age of the women was also a significant factor: for a founder population of ten people, with the women aged between 17 and 21, the probability of extinction was found to be 28 per cent, rising to 77 per cent if the women are aged between 26 and 30. Certainly it is clear that population increase following island colonisation can be very rapid. Pitcairn Island was settled by the Bounty mutineers in 1790: an original founding population of fifteen men, twelve women and one infant girl. Between 1790 and 1810 the population rose at an average annual rate of 3.7 per cent, much higher than the normal figure in a mainland community (Nicholson 1965; Terrell 1986). By 1825 the population had reached (and probably exceeded) the island's carrying capacity, and John Adams, the last surviving mutineer, requested Naval assistance in removing the community to Tasmania. The population was, in fact, moved twice as a result of famine, once to Tahiti in 1831, and once to Norfolk Island in 1835, though in both cases the community returned to Pitcairn. Demography, however, is not an independent variable: it is affected by a number of social institutions, most notably marriage and incest rules. Firth (1961) gives some information on Tikopia, the population of which seems to have been near carrying capacity at the time of his fieldwork. He describes a series of cultural mechanisms to limit population growth, including abortion, infanticide and the widespread and institutionalised practice of coitus interruptus (a technique which young men practised with older, infertile women, before being allowed to establish relationships with younger women). Needless to say, these mechanisms would in most cases be impossible to recognise archaeologically. Population increase, if unchecked, would place severe pressure on the natural environment, and there are numerous examples, particularly in the Pacific, of resource depletion and decreased biodiversity following colonisation by people (Kirch and Yen 1982; Green 1976). Human pressure may be reflected, not only in the decline and ultimate extinction of plant and animal species, but also in evidence for rapid erosion, resulting from clearance and cultivation. This has been noted, for example, on Fiji (Hughes 1985; Rowland and Bent 1982), Vanuatu (Spriggs 1985) and Tikopia (Kirch and Yen 1982). Where a significant human impact on the island ecosystem is identified, however, it is often difficult to establish to what extent this arises from population pressure, and to what extent it is the result of social demands for increased surplus production. On Easter Island, for example, where the human impact on the environment seems to have been catastrophic

(Bahn and Flenley 1992), there is every reason to believe that this was the result of the surplus production engaged in the construction of large stone monuments.

One claim that has frequently been made about island societies (cf. Evans 1973; Vayda and Rappaport 1963; Sahlins 1955) is that they often display a tendency towards the elaboration of particular cultural traits. The most spectacular manifestation of this is on Easter Island (Bahn and Flenley 1992) where large stone platforms (ahu moai) were built, incorporating massive statues. The elaborate initiation rites of Malekula (New Hebrides), involving the construction of stone platforms and megalithic alignments (Deacon 1934; Layard 1942) could also be seen as an example of this phenomenon as, of course, could the megalithic temples of Malta (Stoddart *et al.* 1993). Various explanations have been put forward for this phenomenon. Evans (1973) suggests that: 'The isolation and relative security of island life can allow the continuance of trends which, in a mainland environment, are likely to be inhibited by various extraneous factors long before they reach their logical conclusion.' Sahlins (1955), on the other hand, suggests that, in small ecologically constrained islands, such as Easter Island, communal labour could not be directed into major improvements in subsistence production, and that these efforts were instead channelled into an 'esoteric domain of culture'.

Renfrew (1976) argues that megalithic monuments can be seen as 'territorial markers', and suggests that their appearance along the Atlantic facade of Europe in the Early Neolithic was a response to population increase. The ecological constraints of island life may accentuate the problems of population pressure, so perhaps we are simply dealing with a response to these problems. This hypothesis might work in the case of Easter Island. There are in total, twenty-five 'ahu moai' on Easter Island, and these have a predominantly coastal distribution. Ethnographic information suggests that these platforms were the ceremonial centres for descent groups (or 'mata'), and that the statues represent lineage ancestors (McCall 1979). The coastal distribution corresponds to a widespread Polynesian system of land division, in which each lineage has a strip of land extending inland from the sea, giving each community access to both terrestrial and marine resources. The 'territorial marker' hypothesis, however, would not explain the elaborate megalithic rituals of Malekula. The monuments of Malekula are built in the context of male initiation rites, and have no relationship to the division or control of land. The most elaborate megalithic rituals are those of the coastal islet of Vao, whose people have their cultivated plots, not on the islet itself, but on the adjacent coast of the main island of Malekula (Layard 1942).

What these two Pacific examples have in common, however, is that in both cases the construction and use of megalithic monuments is important in the establishment and reproduction of social relations. In the case of Easter Island, the monuments were built in the context of a highly stratified society, with six social classes (Métraux 1957):

Ariki Mau	Great King of the island
Ariki Paka	Nobility
Ivi Atua	Priests
Matato'a	Warriors
Hiru Moru	Commoners
Kio	Subordinates or slaves

The religion, centred around the stone platforms, was controlled by the Ariki Paka and Ivi Atua, whose stone houses were built in the immediate vicinity of the platforms themselves. It is likely that the monuments played an important role in the articulation and reproduction of the asymmetric social relations between the nobility and priests on the one hand, and the warriors, commoners and slaves on the other.

The megalithic monuments of Malekula were constructed in the context of male initiation rites, controlled by lineage elders. The initiate progresses through a series of grades (the number varies from area to area) starting at adolescence. Each stage involves a payment, made to the elders who control the ritual. In the Maki rituals of Vao (Layard 1942), a whole age-set undergoes the initiation together, and entry to the highest grades is open to any man who lives long enough. These monuments, therefore, are involved in the articulation of two sets of social relationships: between elders and younger men, and between men and women (since women are excluded from the ceremonies and are forbidden from entering the sacred sites). In the related, but rather different, Nimangki rituals of the main island of Malekula (Deacon 1934), men undergo the initiation individually, and entry to the highest grades is restricted to men of particular lineages, so that these rituals are also involved in the reproduction of asymmetric social relations between individual men, and between lineages.

In a general theoretical sense, Meillassoux (1972) has argued that relations of power in tribal societies are often mediated through the control of ritual practice, particularly in relation to initiation rites. Through their control of the sacred knowledge involved in initiation rites, a dominant group is able to control young people's route to adulthood (as socially defined) and to economic independence, and are thus able to make demands on their productive labour. As Bender (1985) argues, this control may also operate through control of socially valued material items (often items acquired through exchange), also considered essential in social discourse (for example, items required for bridewealth presentations). In most mainland communities, the mediation and reproduction of social relations involves *both* the control of initiation rites *and* control over the circulation of socially valued material items (cf. Patton 1993). In the case of a remote island, communities may not be involved in exchange systems, and the control of material items may be far less significant. In such cases, elaborate initiation ceremonies and ritual practices may develop. This is perhaps the real explanation for cultural elaboration on islands, though

30

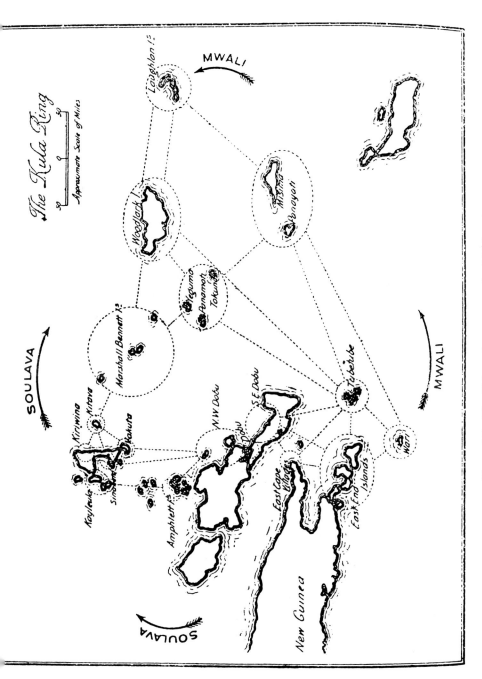

Figure 2.2   The Kula exchange network of Melanesia
Source: Malinowski 1922

it in no way excludes the possibility that the monuments themselves may also serve as 'territorial markers' (as seems to have been the case on Easter Island). This question of cultural elaboration will be explored in Chapter 5 of this book.

Not all island societies are completely isolated from their neighbours on mainlands and on other islands, and the Mediterranean islands in particular are very much less isolated than those in the Pacific. In some cases, island communities are involved in complex and elaborate networks of exchange and interaction, which acquire great social significance. The most famous example of this, and the one most often cited as a general model, is the Kula network of Melanesia, described by Malinowski (1922) and discussed more recently by Leach and Leach (1983), Munn (1986) and Weiner (1976). The Kula network links the Eastern coast of New Guinea with the Amphlett, Trobriand, Dobu and Marshall Bennett archipelagoes, and with the outlying islands of Woodlark, Laughlan, Misima, Sud-Est and Rossel (Figure 2.2). The network involves the circulation of shell arm-rings (mwali) and necklaces (soulava), which move in opposite directions. The exchange of these objects takes place in the context of elaborate rules of reciprocity and etiquette, quite unlike those which govern normal exchange (gimwali).

It can be argued that Kula plays an important role in the mediation of asymmetric relations between elders and younger men, and that the geographical fact of insularity is manipulated by an elite to establish and maintain its own power. Malinowski (1922) stresses that a man's status in Massim society depends upon his involvement in Kula (in most communities, women are excluded from Kula). Scoditti and Leach (1983) state that those outside the Kula network are considered as social marginals (Tanawaga). They also explain how young men become bound by obligations to the elders who initiate them into Kula. Access to the Kula depends upon a network of contacts on other islands, and on the initial possession of Kula valuables with which to enter the exchange system: both are controlled by elders, who are thus able to make demands upon the labour of younger men. Full participation in the Kula depends on the possession of 'Kula magic', which the elders only pass on to the younger men some years after their initial entry to the Kula. Campbell (1983) describes how, in some communities, Kula valuables can enter other spheres of exchange, most notably bridewealth. Munn (1986) makes a similar point with regard to Gawa: here an initial gift of yams and taro are made by the bride's family to the groom's family, and this gift is later reciprocated with a (much more valuable) presentation of Kula valuables and canoes. In such circumstances, the ability of young men to marry, and thus to proceed to economic independence, depends upon their access to Kula, allowing the elders to control the 'means of reproduction' (Meillassoux 1972). The control of access to Kula also operates through the control of canoes. Munn (1986) describes three categories of canoe: Kewo'u, Kalipoulo and Masawa. Of these, it is Masawa, large elaborately carved canoes surrounded by magic and taboo, that are used for Kula. Masawa are not individually owned, but are effectively controlled by a captain or Toliwaga, almost invariably a chief, who decides who sails in the canoe. He controls access, not only to the canoe itself, but also to 'sailing magic', which includes both the ritual and the practical knowledge required to make the long voyages involved in Kula exchange. By controlling men's access to the Kula network, he is able to make demands on them. Malinowski (1922) describes a chief of Omarahana, who controlled a number of canoes, and who was able to requisition 30 to 50 per cent of all the garden produce on Kiriwina, enabling him to finance lavish feasts, and also to support a harem of more than forty concubines.

What is important here is the way in which the geographical fact of insularity is manipulated socially. The control of interaction is not, in itself, a specifically insular phenomenon, but insularity does offer particular possibilities for control, especially through access to boats and knowledge of how to sail them. In these cases, it is exchange, rather than ritual, which becomes elaborated in an insular context. The key variable here is the degree of isolation: in a totally isolated context, as on Easter Island, it is ritual that is most likely to be elaborated, whereas on an island which maintains links with other communities, exchange may become more important. Both can be seen as mechanisms for the expression and reproduction of social relations. There is, of course, no island in the Mediterranean which even approaches the extent of geographical isolation that we see, for example, in Easter Island. Isolation, however, is not simply a factor of geographical circumstances, but also of social and cultural factors. Thus an island community which, at one period in time, is linked to other mainland and island communities by elaborate networks of exchange and interaction, may at another moment in time develop in complete isolation. This seems to have happened, for example, on Malta, where Stoddart *et al.* (1993) link the development of the megalithic temples to the decline of exchange networks with Sicily and southern Italy. The significance of exchange systems in the prehistory of the Mediterranean islands will be considered in Chapter 6.

THE 'ISLAND LABORATORY' RECONSIDERED

This book began with a discussion of John Evans' (1973) idea of the 'island laboratory', an idea which, as we have seen, was originally conceived by Darwin (1968 [1859]), and which was later developed by MacArthur and Wilson (1967) before being applied to archaeology in Evans' seminal paper. Having, in the first two chapters of this book, identified a number of specific characteristics of island, as opposed to mainland societies, we should, perhaps, return to this idea, and ask whether it is really appropriate to use islands as 'laboratories' for the understanding of more general cultural processes? Perhaps, after all, the processes affecting island societies are substantially different from those affecting mainland communities? The answer, perhaps, lies in the fact that the processes affecting insular societies are not *qualitatively* different from those affecting continental societies. The mediation of social asymmetry through elaborate ritual practices and control of exchange networks, for example, are not unique to island communities (Patton 1993), though they may become more intense an an island context. This greater intensity of certain variables, linked to the absence of others, may make these processes more visible and more easily comprehensible in an island context. It may, therefore, be possible for archaeologists to use island studies, much as Darwin (1968 [1859]) and MacArthur and Wilson (1967) did, in the formulation of more general models. The epistemology of a social or cultural science such as archaeology is, of course, quite different from that of the biological sciences which concerned Darwin and his successors, and this will naturally be reflected in the structure of such models. The aim, however, is similar: to use the particularities of island communities as a key to understanding the processes which affect all societies.

3

THE COLONISATION OF THE MEDITERRANEAN ISLANDS

Colonisation is a logical starting point for any consideration of Mediterranean island prehistory. Whilst the colonisation of a continental area may take place gradually, almost imperceptibly, the colonisation of an island requires a specific event, either an accidental or a deliberate sea voyage. The island biogeography model of MacArthur and Wilson (1967) attempts to provide a basis on which to predict the pattern of colonisation of islands by animal and plant species. Since humans are animals, with the same biological needs as other species, it has been argued (cf. Cherry 1984; Keegan and Diamond 1987) that this model can also be used to predict the human colonisation of islands. For the reasons outlined in Chapter 2, however, there is wide-spread agreement among archaeologists that the island biogeography model needs to be adapted in order to take account of the specific features of human, as opposed to non-human colonists. Keegan and Diamond (1987) have summarised the key biogeographical variables which are likely to affect the colonisation of islands by people: these are divided into three categories, described as distance effects, configurational effects and area effects.

1 Distance effects
   a The chances of death during the voyage increase with distance (MacArthur and Wilson 1967).
   b The 'rescue effect': the closer an island is to its parent population, the lower the chances of extinction, since the population may be 'rescued' by fresh immigrants (cf. Brown and Kodric-Brown 1977).
   c The 'commuter effect': islands which are too small to support a self-sustaining population may be occupied if they are within easy distance of another island or a mainland area offering additional resources.

2 Configurational effects
   a The 'stepping-stone effect': the effect of distance may be offset by the existence of stepping-stone islands between a given island and the adjacent mainland.
   b The 'target effect': an island chain, perpendicular to the axis of travel,

offers a larger target than a single island, and is thus more likely to be discovered (cf. Lewis 1972).

3 Area effects

   a A larger island presents a more easily located target.

   b A colonising population is more likely to colonise a larger island, because of the greater variety of resources.

The three main variables identified by Keegan and Diamond, namely distance, configuration and area, are all quantifiable, so that it ought to be possible to test the validity of the island biogeography model empirically. There are, however, a number of complicating factors.

First, as Cherry (1984) suggests, we must distinguish conceptually between *discovery* and *colonisation*. These concepts are not distinguished by MacArthur and Wilson, since the distinction is not relevant to non-human animals: a lizard or a rat 'discovers' an island by accident and either establishes a population or becomes extinct. A human community, on the other hand, may know of the existence of an island, and may visit it periodically without actually colonising it. Discovery and colonisation, moreover, are affected by different variables: in terms of the 'effects' listed above, 1a, 2a, 2b and 3a are relevant to discovery, whereas 1b, 1c and 3b are relevant to colonisation. This situation is further complicated by the fact that discovery and colonisation are in many cases difficult to distinguish from one another archaeologically: there is an understandable (perhaps unavoidable) tendency (cf. Cherry 1981, 1990) to use the earliest dateable material from an island as an indication of the date of colonisation, though in many cases it is not clear that this represents colonisation, as opposed to periodic visits. In any case, a consideration of the evidence for the earliest colonisation of an island will necessarily suffer from the general problems associated with inductive reasoning (Popper 1959). One discovery of an earlier site may completely change the picture: it is for precisely this reason that Cherry (1990) found it necessary to revise his earlier (1981) paper on the earliest colonisation of the Mediterranean islands.

A second issue identified by Cherry (1984) is the need to consider relevant palaeogeographical factors. The coastlines of the Mediterranean have changed significantly since the Pleistocene (Shackleton *et al.* 1984), so that the biogeographical characteristics of islands today (notably their size and distance from mainland) may be very different from those of the Palaeolithic. For example, 18,000 years ago (the time of the last glacial maximum), Formentera was joined to Ibiza, Menorca to Mallorca (Figure 3.1), Sardinia to Corsica, the Maltese and Egadi groups to Sicily, and Elba to the Italian mainland (Figure 3.2). Most of the Cycladic islands formed part of one large island, and many of the North and East Aegean islands formed part of mainland areas (Figure 3.3). By c. 9000 bp, however, coastlines had reached more or less their present configuration, so that palaeogeographical factors are only relevant to a consideration of Palaeolithic and Mesolithic island communities. The palaeogeography of the Mediterranean islands is shown on Figures 3.1 to 3.4.

MEASURING BIOGEOGRAPHICAL VARIABLES

If the island biogeography model is to be tested empirically, we require some means of quantifying the relevant variables. The key variables identified by MacArthur and Wilson (1967) are island size and distance from the mainland. The quantification of these variables may, at first sight, appear relatively unproblematic: island size (measured in square kilometres) could be divided by distance from the mainland (measured in kilometres) to give an area/distance ratio (A/DR). Thus Sardinia, with a surface area of 24,089 km², lying 205 km from the Italian coast, would have an A/DR of 118, whilst Malta, with a surface area of 237 km², 250 km from the mainland, would have an A/DR of 0.9, making it far less likely to be colonised. There are, however, a number of problems with this approach. The most significant problem is that it ignores the configurational properties of the islands concerned. Crete, for example, is 102 km from the mainland, and has an A/DR of 81. Because of

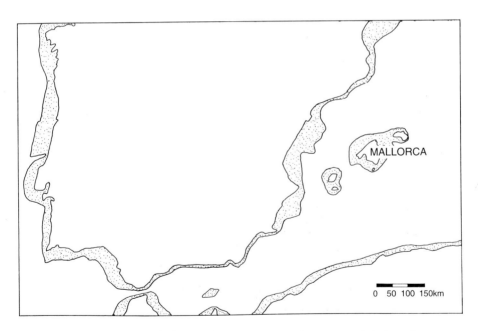

Figure 3.1 Palaeogeography of the western Mediterranean, including the Balearic Islands

Note: The shaded area represents land lost during Holocene sea-level rises

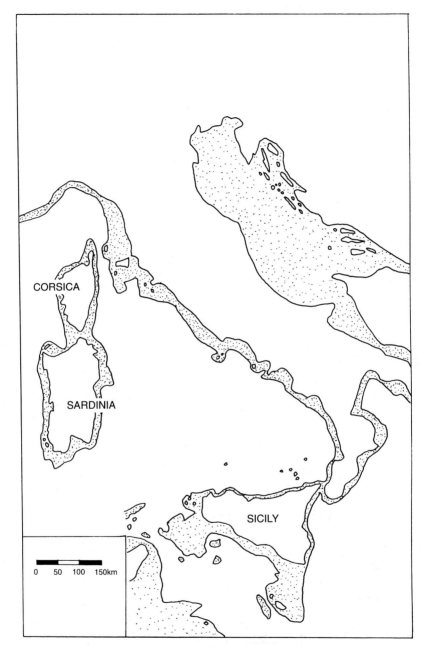

Figure 3.2 Palaeogeography of the central Mediterranean, including Corsica, Sardinia and the Maltese Islands

Note: The shaded area represents land lost during Holocene sea-level rises

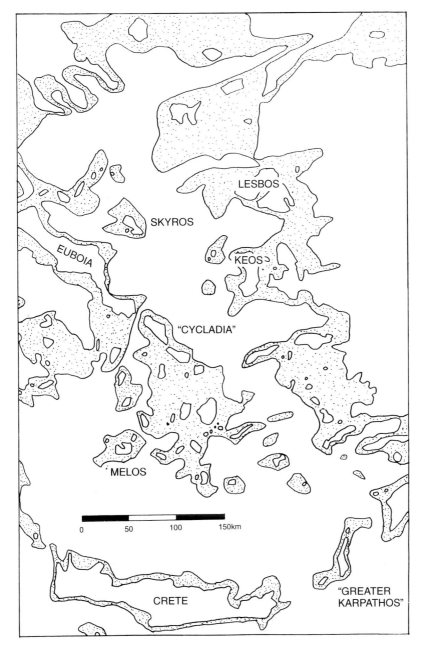

Figure 3.3 Palaeogeography of the Aegean area

Note: The shaded area represents land lost during Holocene sea-level rises

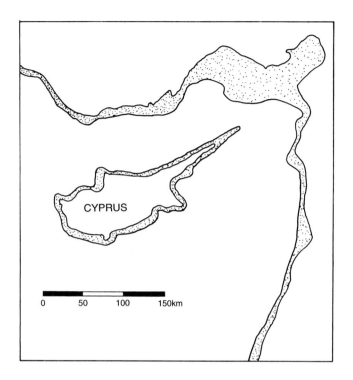

Figure 3.4 Palaeogeography of the eastern Mediterranean, including Cyprus

Note: The shaded area represents land lost during Holocene sea-level rises

the existence of stepping-stone islands, however, the longest single sea voyage that would be required to reach it is 48 km. This figure of the longest single voyage (LSV) seems more meaningful than distance from the mainland, and allows us to create a biogeographic ranking using the following formula:

$$\frac{\text{Area (km}^2)}{\text{Longest Single Voyage (km)}}$$

Using this formula, Crete would have a ranking of 172, rather than 81, whilst Malta would have a ranking of 3, and Sardinia a ranking of 415.

Held (1989) has argued that target width is a more meaningful variable than surface area. Target width is defined as 'The angle which an island subtends on the horizon when viewed from the staging area for a colonisation'. This allows Held to calculate target/distance ratios (T/DR) using the following formula:

$$\frac{\text{Target width (degrees)}}{\text{Distance from staging point (km)}}$$

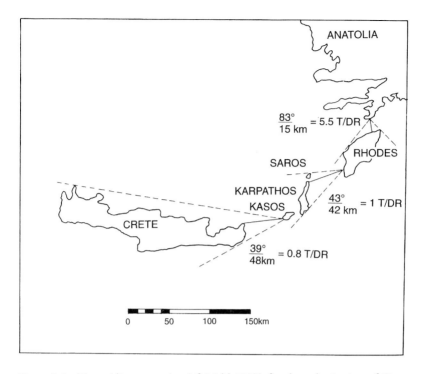

Figure 3.5 Target/distance ratios (cf. Held 1989) for the colonisation of Crete from Anatolia

Note: The islands of Karpathos, Kasos and Saros have been treated as a single group because of the configuration of these islands and the short distances between them

It is important to note that, whereas the biogeographic ranking, as defined above, is an absolute figure, T/DR is relative, in that it always has reference to a specific staging point. This is illustrated by Figure 3.5, which shows the T/DRs involved in the colonisation of Crete by the shortest possible route. The T/DR of Crete from Kasos is 0.8: in order to arrive in Crete, however, a potential colonist has first to reach Rhodes, then Karpathos/Kasos/Saros. If any of these voyages involved a lower T/DR than that between Kasos and Crete, this would have a negative effect on the probability of Crete itself being colonised via this route. We can therefore create a ranking of islands, based on T/DR, by computing the lowest T/DR required to reach an island by the shortest possible route. In the case shown on Figure 3.5, Crete's ranking would thus be 0.8.

Biogeographic ranking and T/DR ranking are not simply different approaches to the same problem: they actually address different issues. T/DR ranking privileges those variables which affect the likelihood that an island will be discovered (1a, 2a, 2b, 3a on pp. 35–36), whereas biogeographic ranking privileges those variables which affect the likelihood that an island will actually be colonised (1b, 1c, 3b above). It may be instructive, therefore, to build two predictive models on the basis of these approaches, and to compare both with the empirically derived data: this should permit an assessment of the relative significance of maritime technology/navigational knowledge and ecological factors as barriers to island colonisation. A close match between T/DR ranking and the empirical data would suggest maritime technology as the main barrier to island colonisation, whereas a close match between biogeographic ranking and empirical data would suggest that ecological factors (i.e. the range of resources available on an island) were more significant.

One further factor which must surely have affected island discovery is visibility. An island which is directly visible from land does not require to be discovered: its existence would have been known of from earliest times, even if people did not have the means to reach it. A second category of island would be one which, although not directly visible from land, could nonetheless be reached without sailing out of sight of land (i.e. it appears on the horizon before the coast of the staging point disappears). The third category of island would be one which could only be reached by sailing out of sight of land (i.e. in navigational terms, it could not be located purely by pilotage). Visibility can be calculated using standard navigational formulae, such as those given in *Inman's Nautical Tables*. A further predictive model for island discovery could be developed on the basis of this criterion, again, focusing on the longest single sea-crossing required to reach a given island.

All of these approaches to quantifying biogeographic variables, and building predictive models on the basis of them, share a common assumption, which cannot pass without comment. This assumption is that colonisation is most likely to have taken place from the closest possible staging point, and by the shortest possible route. In most cases, this assumption is probably a fair one, but there are significant exceptions. The archaeological evidence for the Aceramic Neolithic of Cyprus, for example, suggests affinities with the Northern Levant, rather than with the geographically closer area of southern Anatolia (Todd 1987), whilst Pantelleria and Lampedusa, although closer to North Africa, seem to have been colonised from Sicily. If we take an island such as Cyprus as an example, with long mainland coastlines both to the north and east (Figure 3.4), there are an infinite number of possible staging points along these coasts. If we are to build predictive models which rank islands in terms of their accessibility, we need some means of selecting one of these, and the only logical approach is to select the staging point closest to the island concerned. We can then seek specific explanations for those colonisation events which seem not to have followed this pattern.

BIOGEOGRAPHIC FEATURES OF THE MEDITERRANEAN ISLANDS

Before we can begin to build predictive models, we need to quantify the relevant variables in relation to the islands of the Mediterranean. On the basis of the above discussion, we can identify the following variables as relevant to this consideration.

1   Distance (defined as the longest single sea-crossing required to reach an island).
2   Surface area.
3   T/DR ranking (defined as the lowest T/DR involved in the colonisation of an island).
4   Visibility category: A   Islands directly visible from mainland
                            B   Islands which can be reached without sailing out of sight of land
                            C   Islands which cannot be reached without sailing out of sight of land.

These variables will need to be recorded separately for the Pleistocene and Holocene periods. Strictly speaking, of course, there is no simple Pleistocene or Holocene geography of the Mediterranean region, since the transition from one to the other was a gradual one. For the sake of convenience, however, we will consider the Mediterranean islands at c. 18,000 bp and at c. 9000 bp (see Figures 3.1 to 3.4). The relevant biogeographic features of the Mediterranean islands during these two periods are shown on Tables 3.1 and 3.2.

THREE PREDICTIVE MODELS

On the basis of the biogeographical data summarised in Tables 3.1 and 3.2, we can formulate three predictive models for the colonisation of the Mediterranean islands. These models will be based on visibility criteria, target/distance ratios and biogeographic ranking, respectively. These models can then be compared with the patterns identified from the empirical data (cf. Cherry 1981, 1990).

Visibility model

This model focuses on discovery, rather than on colonisation, and is based on the simple assumption that the first islands to be 'discovered' will be those which are directly visible from the mainland (category A on Table 3.2). Given that navigation is likely to have depended on pilotage, we might expect that the next group of islands to be discovered would be those in Category B (islands which can be located without sailing out of sight of land) and that islands in Category C would be the last to be discovered.

Table 3.1 Biogeographic features of the Mediterranean islands at *c.* 8000 bp

Island	Distance	Surface area	T/DR ranking	Visibility category
Greater Ibiza (1)	42	4756	1.1	B
Greater Mallorca (2)	42	14560	1.1	B
'Corsardinia' (3)	9	45000	19.4	A
Stromboli	20	13	1.8	B
Greater Lipari (4)	20	60	1.8	A
Salina	20	26	1.8	B
Filicudi	20	10	1.5	B
Alicudi	20	5	1.5	B
Ustica	41	8	0.3	A
Pantelleria	8	220	7.4	A
Linosa	19	240	3.4	A
Greater Ithaca (5)	2	4200	2	A
Crete	2	13000	72.5	A
Gavdhos	19	525	3.4	B
Melos	5	1800	12.2	B
Greater Kythnos (6)	2	1260	45	A
'Cycladia' (7)	5	17680	26.2	A
Anaphi	8	625	13.3	B
Astipalaia	7	660	7.7	B
Greater Karpathos (8)	10	1200	8	B
Rhodes	1	3150	70	A
Tilos	3	360	3.2	A
Psara	3	684	22	A
Skyros	7	1440	10	A
Cyprus	30	15000	1.8	A

Note: (1) Ibiza plus Formentera; (2) Mallorca plus Menorca; (3) Corsica plus Sardinia; (4) Lipari plus Vulcano; (5) Ithaka, Kephallinia and Zakynthos; (6) Kythnos plus Serifos; (7) Andros, Tinos, Mykonos, Paros, Naxos, Amorgos, Ios, Sikinos and Thera; (8) Karpathos plus Kasos and Saros.

Using these criteria we would expect 'Corsardinia', Greater Lipari, Ustica, Pantelleria, Linosa, Crete, Greater Kithnos, 'Cycladia', Rhodes, Tilos, Psara, Skiros and Cyprus to be the islands most likely to be colonised during the Pleistocene (Table 3.1).

For the Holocene, the islands are grouped as follows (Table 3.2).

Category A:
- Atokos, Corfu, Kalamos, Kephallenia, Kythera, Lefkas, Meganissi and Zakynthos in the Ionian group.
- The islands of the Argo-Saronic group.
- Samothrace and Thassos in the North Aegean.
- Skiathos, Skopelos and Euboia in the Northern Sporades.
- Chios, Lesbos and Samos in the East Aegean.
- Kalymnos, Kos, Rhodes and Syme in the Dodecanese.
- Andros, Keos and Makronissos in the Cycladic group.
- Sicily, the Egadi Islands, the Pontine Islands, Ischia, Capri,

Elba, Giglio, Giannutri, Lipari, Salina, Vulcano, Stromboli, Filicudi and Alicudi in the Tyrhennian group.

- The Tremiti Islands in the Italian Adriatic.

Category B:
- Andikythera, Arkoudi and Ithaca in the Ionian group.
- Lemnos in the North Aegean.
- Halonissos, Kyra Panagia and Skyros in the Northern Sporades.
- Ikaria and Psara in the East Aegean.
- All of the Dodecanese and Cycladic Islands, other than those listed above.
- Crete, Dia and Gavdhos in the South Aegean.
- Cyprus, Malta, Gozo and Comino.
- Corsica, Sardinia, Ustica, Montecristo and Panarea in the Tyrhennian group.
- Pianosa in the Italian Adriatic.
- All of the Balearic Islands.

Category C:
- The islands of Pantelleria and Lampedusa in the central Mediterranean.

Target/distance ratio model

Like the first model, this focuses on discovery, rather than colonisation. The assumption in this case is that the islands with the highest T/DR ranking are most likely to be colonised first, whereas those with the lowest rankings are the least likely to be colonised. Thus for the Pleistocene, Rhodes and Crete are the most likely islands to be colonised (Table 3.1), whilst for the Holocene, islands such as Euboia, Salamis and Lefkas might be expected to yield the earliest evidence for human activity. In general, we would expect the relationship between T/DR ranking and the date of earliest human activity to be a linear one (Figure 3.6(a)).

Biogeographic ranking model

This model focuses on colonisation rather than discovery, and assumes that the size of an island (considered as an indication of its resources and 'carrying capacity') is an important variable. Distance, however, is also an important factor, because of the 'rescue effect' and the 'commuter effect', as outlined above. The assumption, therefore, is that the islands with the highest biogeographic rankings are those most likely to be colonised first. We would thus expect to find a linear relationship between biogeographic ranking and date of earliest human activity (Figure 3.6(b)).

In the case of the second and third models, we would not, of course, expect to see a precise linear relationship between variables, though we would expect to see a general linear trend.

Table 3.2 Biogeographic features of the Mediterranean islands at *c.* 9000 bp

Island	Distance	Surface area	T/DR ranking	Visibility category
Ionian Islands				
Andikythera	35	20	0.1	B
Arkoudi	5	5	6.2	B
Atokos	8	5	2.5	A
Corfu	5	593	27.2	A
Ithaca	9	96	3.3	B
Kalamos	2	25	30	A
Kephallenia	9	781	5	A
Kythera	15	280	2.7	A
Lefkas	0.5	303	210	A
Meganissi	1	20	61	A
Zakynthos	18	402	2.8	A
Argo-Saronic Islands				
Aegina	21	83	2.4	A
Hydra	6	50	14.7	A
Poros	0.5	23	192	A
Salamis	0.5	96	200	A
Spetsai	2	22	35	A
North Aegean Islands				
Lemnos	28	478	1.8	B
Samothrace	25	178	0.8	A
Thassos	7	380	9	A
Northern Sporades				
Halonissos	10	65	5	B
Kyra Panagia	10	25	3	B
Skiathos	4	50	15.5	A
Skopelos	10	97	6.3	A
Skyros	33	210	0.8	B
Euboia	0.5	3645	360	A
East Aegean Islands				
Chios	11	842	10	A
Ikaria	18	256	8.6	B
Lesbos	12	1633	7.2	A
Psara	19	40	1.3	B
Samos	5	477	26	A
Dodecanese Islands				
Alimnia	19	7	4.5	B
Astipalaia	48	97	0.4	B
Castellorizo	10	10	3.8	B
Chalki	10	28	3.8	B
Giali	10	9	3.5	B
Kalymnos	5	93	4.6	A
Karpathos	48	301	1	B
Kasos	48	69	1	B
Kos	5	290	16.2	A
Leros	5	53	4.6	B

Island	Distance	Surface area	T/DR ranking	Visibility category
Lipsoi	9	17	3.8	B
Nisyros	11	37	3.5	B
Patmos	9	34	3.8	B
Rhodes	15	1400	5.5	A
Saros	48	21	1	B
Syme	8	38	4	A
Cycladic Islands				
Amorgos	20	124	1.2	B
Anaphi	25	40	0.7	B
Andros	10	380	4.9	A
Antiparos	18	35	1.2	B
Delos	0.5	3	4.9	B
Despotiko	20	8	1.2	B
Donoussa	20	14	0.8	B
Heraklia	20	18	1.2	B
Ios	20	109	1.2	B
Keos	12	131	4	A
Keros	20	15	1.2	B
Kimolos	20	36	2.1	B
Kouphounissia	20	6	1.2	B
Kythnos	15	100	4	B
Makronissos	3	18	29.3	A
Melos	20	151	2.1	B
Mykonos	11	86	4.9	B
Naxos	20	430	1.2	B
Paros	20	196	1.2	B
Rheneia	11	14	4.9	B
Pholegandros	20	32	1.2	B
Schinoussa	20	9	1.2	B
Serifos	20	75	4	B
Sikinos	20	43	1.2	B
Siphnos	20	74	2.8	B
Syros	15	85	3.1	B
Thera	20	76	1.2	B
Therassia	20	9	1.2	B
Tinos	12	195	4.9	B
South Aegean Islands				
Crete	48	8259	0.8	B
Dia	48	12	0.7	B
Gavdhos	48	30	0.4	B
East Mediterranean Islands				
Cyprus	60	9251	1.7[1]	B
Maltese Islands				
Malta	80	237	0.1	B
Gozo	80	66	0.1	B
Comino	80	3	0.1	B
Pantelleria	72	82	0.1[2]	C
Lampedusa	130	20	0.03[3]	C

Island	Distance	Surface area	T/DR ranking	Visibility category
Tyrhennian Islands				
Sicily	3	25708	56.3	A
Egadi Islands	38	13	1.3	A
Sardinia	58	24089	1.5	B
Corsica	58	8722	1.5	B
Pontine Islands	32	12	1	A
Ischia	11	46	2.7	A
Capri	5	10	6	A
Ustica	55	8	0.07	B
Elba	10	220	5.2	A
Montecristo	30	10	0.3	B
Giglio	15	15	1.8	A
Giannutri	14	3	0.7	A
Lipari	20	38	1.5	A
Salina	20	26	1.5	A
Vulcano	20	20	1.5	A
Panarea	20	3	0.4	B
Stromboli	20	13	0.4	A
Filicudi	20	10	0.8	A
Alicudi	20	5	0.7	A
Italian Adriatic Islands				
Tremiti Islands	25	6	0.6	A
Pianosa	25	1	0.4	B
Balearic Islands				
Ibiza	92	541	0.2	B
Mallorca	92	3740	0.2	B
Menorca	92	702	0.2	B
Formentera	95	82	0.2	B
Cabrera	92	13	0.2	B
Conejera	92	18	0.2	B

Note: 1 This T/DR rating is calculated from the closest point on the Anatolian coast. The archaeological evidence suggests, however, that the origins of the Cypriot Aceramic Neolithic are to be found in the northern Levant. The lowest T/DR rating from this area is 0.3 (from Ras Ibn Hani). 2 This T/DR rating is calculated from the closest point on the North African coast. The archaeological evidence, however, suggests that Pantelleria was colonised from Sicily, in which case the T/DR rating is 0.06. 3 Lampedusa, like Pantelleria, seems to have been colonised from the northern, rather than the southern shore of the Mediterranean. From its closest 'stepping stone', Malta, the T/DR rating is 0.01.

THE PREHISTORIC COLONISATION OF THE MEDITERRANEAN ISLANDS: THE EMPIRICAL DATA

In order to assess the predictive models outlined above, we need to return to the empirical data. These have been admirably summarised by Cherry (1981, 1990) and there is little to add here, except to update some points.

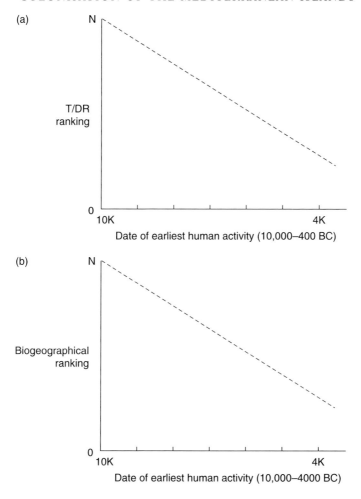

Figure 3.6 Predicted patterns of island colonisation in Mediterranean prehistory based on (a) target/distance ratio model and (b) biogeographic ranking model

Ionian Islands

The most recent significant discovery (Kavvadias 1984) is of Middle Palaeolithic evidence at Nea Skala on Kephallinia (part of 'Greater Ithaca'). As Cherry (1990) points out, this is particularly important, since it does indicate Palaeolithic activity on a true island. There is, however, no evidence for continuation of this activity into the Upper Palaeolithic. Meganissi and Lefkas have produced possible evidence for activity in the fourth millennium cal. BC, whilst Kythera, Ithaca and Kephallinia seem to have been occupied in the third millennium. Zakynthos was occupied in the second millennium cal. BC, whilst the remaining Ionian Islands (Andikythera, Arkoudi, Atokos, Kalamos) have produced no evidence for activity prior to the first millennium (Cherry 1984).

Argo-Saronic Islands

Despite the accessibility of these islands, the earliest evidence for human activity dates to the fourth millennium cal. BC on Aegina, and to the third millennium on the other islands (Cherry 1981).

North Aegean Islands

The islands of Thasos, Lemnos and Samothrace have produced evidence for activity during the fifth millennium cal. BC (Catling 1986; Matsas 1984, 1987). Palaeolithic evidence from Thasos is irrelevant, since Thasos would not have been an island during the Pleistocene (Cherry 1990).

Northern Sporades

Efstratiou (1985) has published evidence for Palaeolithic activity on Halonissos, Mikro Kokkinokastro, Kyra Panagia and Skyros. Cherry (1990) questions whether these would in fact have been islands during the Pleistocene, though in fairness, Skyros probably was (see Figure 3.3; Table 3.1). The site of Aghios Petros, on Kyra Panagia, demonstrates human activity during the sixth millennium cal. BC (Efstratiou *op. cit.*).

East Aegean Islands

The islands of Chios, Samos and Psara have produced evidence for activity during the fourth millennium cal. BC (Cherry 1984, 1990). There is, surprisingly, no evidence for occupation on Lesbos prior to the third millennium, or on Ikaria prior to the first millennium.

Dodecanese Islands

Material dating to the fourth millennium cal. BC has been identified on Astipalaia, Giali, Kalymnos, Karpathos, Kasos, Kos, Leros, Saros and Syme (Cherry 1990). Nisyros seems to have been occupied in the third millennium and Patmos and Lipsoi in the second. Alimnia, Castellorizo and Chalki have, as yet, produced no evidence for human activity prior to the first millennium.

Cycladic Islands

Sites of the 'Saliagos culture' on Paros, Antiparos and Mykonos can be dated to the sixth/fifth millennia cal. BC (Evans and Renfrew 1968). Keos, Makronissos, Melos, Naxos and Thera have produced fourth-millennium sites

(Hadjianastasiou 1988; Sotirakopoulou 1989). Cherry (1990), however, has drawn attention to the great number of Cycladic islands which were apparently occupied for the first time during the third millennium (Amorgos, Andros, Delos, Despotiko, Donoussa, Heraklia, Ios, Keros, Kimolos, Kouphounissia, Kythnos, Paros, Rheneia, Schinoussa, Sikinos, Siphnos, Syros and Tinos). Other Cycladic islands have produced no evidence for activity prior to the second (Anaphi, Serifos, Therassia) or first (Pholegandros) millennia.

South Aegean Islands

The earliest deposits at Knossos (Crete) have produced a series of important radiocarbon dates:

8050 ± 180 bp = 7160–6660 cal. BC (BM-124)
7910 ± 140 bp = 7020–6500 cal. BC (BM-278)
7740 ± 140 bp = 6680–6420 cal. BC (BM-436)
7570 ± 150 bp = 6480–6200 cal. BC (BM-272)
7000 ± 180 bp = 6050–5800 cal. BC (BM-126)

East Mediterranean Islands

The site of Akrotiri Aetokremnos (Simmons 1991), on Cyprus, has produced a series of early radiocarbon dates, ranging from 10,340 ± 130 bp to 9040 ± 160 bp (this evidence is discussed in more detail in the following chapter). This seems, however, to represent a relatively short-lived occupation, and there is certainly no evidence for continuity between the hunter–gatherer communities represented at Akrotiri, and the Aceramic Neolithic communities represented at Kalavassos-Tenta (Todd 1987), where the radiocarbon dates cluster in the mid-ninth millennium bp. Watkins (1981) has identified a similar 'gaping chasm' between the Aceramic and Ceramic Neolithic phases on Cyprus, suggesting that the island may have been abandoned and then recolonised on at least two occasions.

The Maltese Islands

The earliest dated material from Malta is from Ghar Dalam, with a radiocarbon date calibrating between 5260 and 4840 BC (Cherry 1981). Gozo has produced pottery similar to that found at Ghar Dalam. The earliest known material from Comino is of Roman age. The earliest direct evidence for human occupation on Pantelleria is from the Bronze Age site of La Mursia, dating to the second millennium cal. BC. Pantellerian obsidian, however, is found on the earliest Maltese sites, demonstrating that Pantelleria (one of the most remote islands in the Mediterranean) had at least been discovered (if not actually settled) by the sixth millennium cal. BC. Sites on Lampedusa have produced Stentinello-type pottery, dating to the sixth millennium.

Tyrhennian Islands

Significant Palaeolithic evidence is known both from Sicily and from the Egadi Islands, but this is not relevant to a consideration of island colonisation, since these would not have been islands during the Pleistocene (Cherry 1990). The colonisation of Sardinia has become a matter of considerable controversy in recent years. Claims have been made for Lower Palaeolithic occupation on the basis of finds from Dorgali (Blanc 1955) and Anglona (Arca *et al.* 1982). Cherry (1990, 1992) urges caution over these claims which, although plausible, are based on typological arguments without any corroborating palaeontological or stratigraphic evidence. More significant, however, is the evidence from the lowest level at the Corbeddu Cave (Sondaar *et al.* 1984), where remains of the endemic deer *Megaceros cazioti* produced radiocarbon dates of 13,590 ± 140 bp (GRN-11435) and 14,370 ± 190 bp (UtC-242). Human influence on the faunal assemblage has been claimed both in the horizontal distribution of the remains and in the existence of butchery marks, but no artefacts were recovered and Cherry (1992) disputes the claims of human activity. An overlying layer (with radiocarbon dates between 7860 bp and 11,040 bp) produced human remains which, according to Sondaar *et al.* (1984), show endemic features, suggesting a long period of insular development (this evidence is discussed in more detail in the following chapter). At this stage we should probably treat the arguments for a long period of Pleistocene human occupation with considerable caution (Vigne 1989; Cherry 1992), and even the relatively late evidence from the lower level at Corbeddu must be seen as inconclusive. The evidence for human activity in the overlying layer at Corbeddu, however, is incontestable, demonstrating a human presence at a relatively early stage in the Holocene. On Corsica, the earliest evidence is from Strette (Vigne *et al.* 1981), where a horizon with animal bones and a lithic assemblage produced a radiocarbon date of 9140 ± 300 bp (LY-2837).

Lipari seems to have been colonised early in the Neolithic: Stentinello pottery (sixth millennium cal. BC) has been found at Castellaro Vecchio and Lipari Acropolis, and Lipari obsidian is found in the earliest cultural horizons on Malta. Chalcolithic material (fourth millennium cal. BC) has been found on Salina, Panarea, Stromboli, Filicudi, Palmarola, Ischia and Capri (Cherry 1981).

Italian Adriatic Islands

Sites on the Tremiti Islands have produced Early Neolithic pottery, dating to the sixth millennium cal. BC (Cherry 1981), whilst Pianosa shows no evidence of having been occupied prior to the Chalcolithic (fourth millennium cal. BC).

Balearic Islands

The earliest evidence for human activity on the Balearic Islands, characterised by Waldren (1982) as the 'Early Settlement Horizon' is dated by radiocarbon dates from Son Muleta (5935 ± 109 bp: KBN-640d) and Son Muleta (6680 ± 120 bp: QL-29). Material of this period, however, has as yet been found only on Mallorca. The earliest evidence for human activity on Ibiza, Menorca and Formentera dates to the third millennium cal. BC. Cabrera and Conejera have produced no evidence for occupation prior to the Punic period (first millennium cal. BC).

TESTING THE PREDICTIVE MODELS

Following this brief review of the material evidence, we are now in a position to test the three predictive models outlined above.

Visibility model

Looking initially at the evidence for the Pleistocene, twenty-five islands are considered on Table 3.1. Of these, fourteen fall into category A (islands directly visible from land), and the remaining eleven into category B (islands which can be reached without sailing out of sight of land). Three of these islands, 'Greater Ithaca', 'Corsardinia' and Skyros, have produced possible evidence for human activity during the Pleistocene. All of these islands fall into category A (in fact all are situated less than 10 km from the mainland), confirming the expected pattern.

The Holocene pattern, however, is rather more complicated. Of the ninety-five islands which have produced dateable evidence for prehistoric activity in the Holocene (Table 3.2), thirty-four fall into category A, fifty-nine into category B and two into category C. The pattern of colonisation of islands within these three categories can perhaps best be illustrated by means of a cumulative frequency diagram (Figure 3.7). It can be clearly seen that the pattern derived from the data does not correspond with that predicted by the visibility model: in fact, the first islands to be colonised during the Holocene appear predominantly to fall into category B (Cyprus, Corsica, Sardinia, Crete), rather than category A. Only in the fourth millennium cal. BC does the rate of colonisation of category A islands overtake that of category B islands, and then only marginally. Even more surprising is the early date of the evidence from the two category C islands (Pantelleria and Lampedusa), which seem to have been exploited well in advance of many category A islands. To some extent these anomalies could simply reflect the incomplete nature of the data: perhaps there is a great deal of early evidence yet to be discovered on some of the category A islands. It seems difficult, however, to account for such a striking pattern purely in terms of recovery

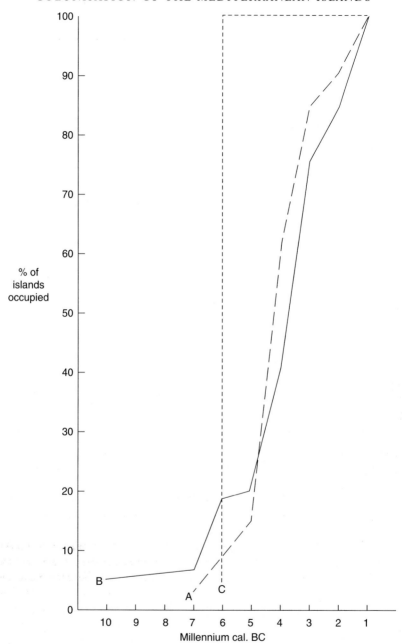

Figure 3.7   Cumulative frequency diagram showing the pattern of colonisation of islands in visibility categories A, B and C

bias. It is equally difficult to believe that the pattern shown on Figure 3.7 reflects the genuine pattern of *discovery* of the Mediterranean islands during the Holocene, since it would be difficult to reach most of the category B and C islands without passing close by other islands in category A. This is well illustrated by the case of Crete, which was presumably colonised either from southwest Anatolia, or from southeast Greece. It would be difficult to imagine Anatolian colonists discovering Crete without noticing Rhodes, Saros, Karpathos or Kasos on the way (Figure 3.5), or Greek colonists arriving in Crete without noticing Kythera or Andikythera. Crete, however, was colonised by the seventh millennium cal. BC, whereas none of the stepping-stone islands show any evidence for activity prior to the fourth millennium. The clear implication of this is that the rate of *colonisation* does not reflect the rate of *discovery*, and that maritime technology and navigational skill were not the main barriers to island colonisation in the Holocene. Given the early evidence from some of the more remote islands, it seems likely that most Mediterranean islands had been discovered at least by the sixth millennium cal. BC: the fact that many of them seem not to have been colonised until much later suggests that the availability of resources may have been a more significant factor in determining colonisation patterns. This might also explain the remarkably early evidence for human activity on Pantelleria, one of the most remote islands which, significantly, is also an important source of obsidian.

Target/distance ratio model

The three islands which have produced possible evidence of Pleistocene human activity ('Corsardinia', 'Greater Ithaca' and Skyros) all have relatively high T/DR rankings (19.4, 25.5 and 10, respectively), though significantly these are not the highest rankings. Rhodes and Crete, for example, have T/DR rankings of 70 and 72.5, respectively, 'Greater Kythnos' has a ranking of 45, yet none of these islands has provided any convincing evidence for Pleistocene human activity. It is likely, in any case, that Pleistocene island colonisation was a largely chance affair.

Figure 3.8 shows the relationship between T/DR ranking and earliest evidence for Holocene activity on the Mediterranean islands. These data clearly do not show the predicted linear trend: the islands to be colonised earliest have low T/DR rankings (Cyprus, Sardinia, Corsica), whilst there are a number of islands with very high T/DR rankings which were apparently not colonised until the third millennium cal. BC (Lefkas, Salamis, Poros). T/DR Ranking, as we have already seen, reflects the probability of *discovery* rather than *colonisation*, so that these data corroborate those relating to the visibility model, suggesting that variables affecting discovery are not the most relevant ones to an attempt to model the Holocene colonisation of the Mediterranean islands.

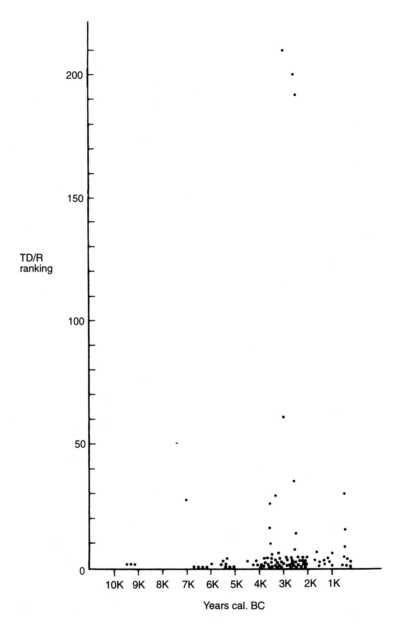

Figure 3.8 Scatter plot showing the relationship between T/DR ranking and earliest evidence for Holocene human activity on the Mediterranean islands

Biogeographic ranking model

The biogeographic rankings of the islands identified for the Pleistocene (Table 3.1) range from 0.2 (Ustica) to 6500 (Crete). The three islands which have produced evidence for Pleistocene human activity all have relatively high rankings ('Corsardinia' –5000; 'Greater Ithaca' –2100; Skyros –205.7), though there are other islands with higher rankings which have produced no such evidence (Crete –6500; Rhodes –3150; 'Cycladia' –1536). Once again, this may simply reflect the chance element in Pleistocene human colonisation, though it is unlikely that islands with very low rankings could have supported Pleistocene human populations.

Figure 3.9 shows the relationship between biogeographic ranking and earliest evidence for Holocene human activity on the Mediterranean islands. These data do show a generalised linear trend, though there are two significant groups of outliers. First, there is a group of islands with very low rankings which were colonised surprisingly early, in the sixth millennium cal. BC. These include Kyra Panagia in the Northern Sporades, Antiparos, Mykonos and Paros in the Cyclades, and the islands of Malta, Gozo, Pantelleria, Lampedusa and Lipari in the central Mediterranean. Second, there is a group of islands with high rankings which were colonised surprisingly late (third/fourth millennia cal. BC), including Lefkas in the Ionian Islands and Salamis in the Argo-Saronic Islands. Once again, it is difficult to believe in this as a reflection of discovery rates: Salamis is only 0.5 km from the mainland, and cannot possibly have remained undiscovered until the third millennium cal. BC. The fact that the biogeographic ranking model provides a better prediction of Holocene island colonisation patterns than the other two models suggests that it was primarily resource availability rather than maritime knowledge and technology that determined whether or not islands were settled and exploited. This may also help us to explain some of the anomalies on Figure 3.9. The islands with low rankings which were occupied (or at least exploited) during the sixth millennium cal. BC include Paros, Antiparos and Mykonos (where sites of the 'Saliagos Culture' seem to have been involved in a specialised fishing economy (Evans and Renfrew 1968; Bintliff 1977), and the central Mediterranean islands of Pantelleria and Lipari (which were major sources of obsidian). Biogeographic ranking is calculated from surface area (taken as an indication of the terrestrial biodiversity of an island) and distance (taken as an indication of the operation of the 'rescue' and 'commuter' effects). Permanent settled communities with a subsistence economy based on terrestrial resources would thus be expected to favour islands with a high ranking. Communities with more specialised economies, however, would select islands for colonisation according to a different set of criteria (the availability of specific biological and/or mineral resources) and this is precisely what seems to have happened. The relationship between resource availability and island settlement will be considered in more detail in Chapter 4.

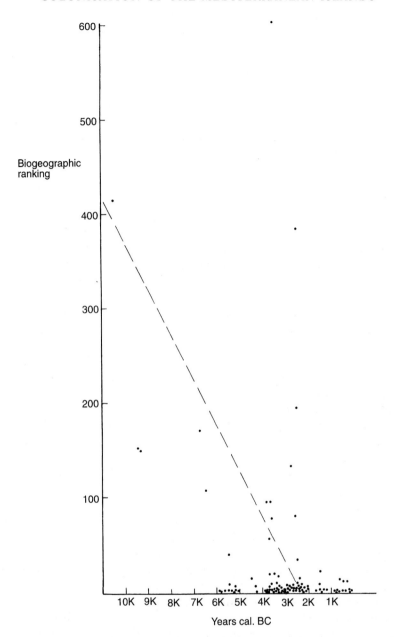

Figure 3.9 Scatter plot showing the relationship between biogeographic ranking and earliest evidence for Holocene human activity on the Mediterranean islands

PATTERNS OF ISLAND COLONISATION IN MEDITERRANEAN PREHISTORY

The archaeological evidence outlined above suggests a marked difference between patterns of island colonisation during the Pleistocene and Holocene periods. During the Pleistocene, very few islands seem to have been occupied by human communities. Those which were exploited were all relatively accessible (directly visible from the mainland, with high T/DR rankings and biogeographic rankings). There are many islands, however, which fulfil all of these criteria, yet seem not to have been occupied, suggesting, perhaps, a significant element of chance in the process of colonisation.

The Holocene pattern is very different. Many of the islands which were settled at an early stage in the Holocene are relatively inaccessible (not directly visible from land, and with low T/DR Rankings): the most remote islands in the Mediterranean (Lampedusa and Pantelleria) had certainly been discovered by the sixth millennium cal. BC. This suggests a revolution in maritime technology and navigational knowledge at a relatively early stage in the Holocene. To some extent this may reflect an adaptation to changing environmental conditions: as sea-levels rose and extensive land areas were lost, Mediterranean communities may have become increasingly dependent upon the resources of the sea, leading to the development of more sophisticated sea-craft, and to the accumulation of navigational knowledge. This, however, may be only one part of the picture: the adoption of specialised economic strategies may also reflect an intensification of production for social reasons, as suggested by Bender (1978).

Figure 3.10 shows the overall chronological pattern of Mediterranean island colonisation during the Holocene, based on the evidence available at the time of writing. Only three islands have produced evidence for human activity in the Early Holocene. Corsica, Sardinia and Cyprus are three of the largest islands in the Mediterranean (Figure 1.1), and the range of available resources may have permitted these islands to support permanent hunter–gatherer populations. During the later Holocene, three main phases of island colonisation can be identified: the first culminating in the sixth millennium cal. BC, the second in the fourth and third millennia, and the third in the first millennium cal. BC (Figure 3.10).

The first apparent wave of island colonisations, beginning in the seventh millennium and culminating in the sixth millennium cal. BC saw the colonisation of Corfu in the Ionian Islands, Kyra Panagia in the Northern Sporades, Antiparos, Paros and Mykonos in the Cyclades, Crete in the Southern Aegean, Malta, Gozo, Pantelleria, Lampedusa and Lipari in the central Mediterranean and Mallorca in the Balearic Islands. Chronologically, this coincides with the transition to food producing economies across most of the Mediterranean region, increasing the potential carrying capacities of the islands concerned (see Chapter 4), and permitting the colonisation of small islands (Lampedusa

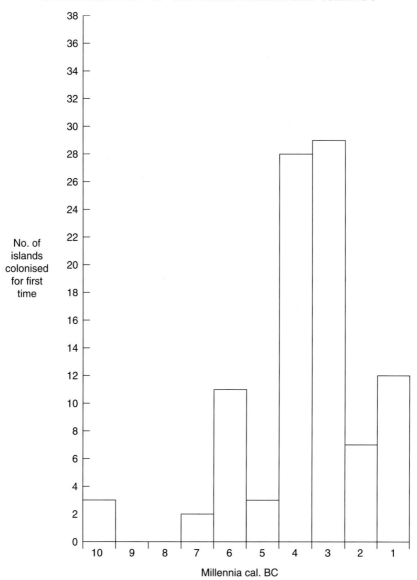

Figure 3.10 Histogram showing the chronology of Holocene island colonisation in the Mediterranean

has a surface area of only 20 km²) which could probably never have supported permanent hunter–gatherer populations. The seventh and sixth millennia cal. BC, however, were also periods of significant social change marked, for example, by the emergence of long-distance exchange networks in the central Mediterranean (see Chapter 6). Increasing social demands for surplus production are likely to have played a significant role in encouraging island colonisation, and the economic specialisation which seems, in some cases, to have been associated with it (intensive fishing in the Cyclades, obsidian extraction on Lipari and Pantelleria).

The second, and most significant, phase of colonisation took place in the fourth and third millennia cal. BC, and was marked by the colonisation of the Argo-Saronic Islands, most of the Dodecanese and Cycladic Islands and most of the remaining islands in the central Mediterranean and in the Balearic archipelago. This phase coincides chronologically with the 'secondary products revolution' identified by Sherratt (1979). Much of the Mediterranean region is dominated by poor, thin soils, unsuitable for large-scale cereal production. In such environments the raising of sheep and goats, and especially the exploitation of their secondary products, has traditionally been an important economic activity: sheep and goats, unlike cattle, can thrive in areas of very poor grazing. Halstead (1981) has stressed the importance of ovicaprids in the development of 'Mediterranean polyculture' (originally defined by Renfrew (1972) as cereals, olives and vine). Like the transition to food production, therefore, the 'secondary products revolution' may have raised the potential carrying capacity of many of the Mediterranean islands, permitting this further wave of colonisation. Like the transition to food production, however, the 'secondary products revolution' cannot be understood except in its social context. This is most clearly seen in the Aegean, where the fourth and third millennia are marked by evidence for increasing social differentiation, with the appearance of fortified citadels as, for example, at Kastri, and of urban centres such as Phylakopi (Renfrew 1972). In Malta this period is marked by the construction of the largest and most spectacular stone temples (see Chapter 7), whilst in Iberia we see the appearance of large concentrations of megalithic monuments (Leisner and Leisner 1943). All of these phenomena would have required increasing surplus production, the demand for which may have been a significant factor in encouraging the exploitation of secondary products. At the same time, the increasingly competitive social relations associated with these phenomena may have encouraged island colonisation as an outlet for the ambitions of defeated rivals.

The final phase of island colonisation, during the first millennium cal. BC, saw the colonisation of virtually all the remaining islands in the Mediterranean. Some of these islands are very small indeed (e.g. Arkoudi and Atokos in the Ionian Islands with surface areas of 5 and 8 km², respectively), and could probably never have supported a self-sufficient community. This period, however, was marked by the development of state societies in Greece, Italy and the Levant, linked to an explosion of international trade which brought about the integration of the economies of widely dispersed areas within the Mediterranean region. This explosion of trade must surely have been one of the factors in promoting island colonisation. Negbi (1992), for example, stresses the importance of island colonies to the development of the Phoenician trade network. The Balearic islands of Cabrera and Conejera were apparently occupied for the first time during the Punic period. Islands were attractive to trading nations for a variety of reasons: they are relatively easy to defend and are, in many cases, strategically located, providing convenient stop-off points on long voyages. State formation also involved periods of intense competition and political conflict, and once again, it is likely that island colonisation became an attractive option for defeated rivals.

In conclusion, therefore, the archaeological evidence for island colonisation in the Mediterranean does not suggest a gradual and continuous process of 'infilling' linked to chance discoveries, but rather a 'punctuated equilibrium' with major phases of colonisation linked to significant economic developments (the 'Neolithic revolution', the 'secondary products revolution' and the trade explosion of the first millennium BC) and to periods of marked social change.

4

INSULARITY AND HUMAN ECOLOGY

The concept of 'island biogeography', from which much of the recent interest in island archaeology has stemmed, is essentially an ecological one, and much of the archaeological literature on island societies has focused on the ecological dimension. In the previous chapter, we identified an important distinction between *discovery* of an island and *colonisation* (processes which have often been confused, and which may be very difficult to differentiate on the basis of archaeological evidence). The variables which influence the probability of discovery include the distance of the island from the mainland, the presence or absence of 'stepping-stone' islands, the size of the island, the level of maritime technology and the nature and extent of maritime activity. The probability of colonisation, on the other hand, is determined primarily by ecological factors, notably the island's carrying capacity. Carrying capacity is defined by Williamson and Sabath (1984) as a 'potential to sustain a certain equilibrium human population size ... ranging from 0 to many thousands'.

This carrying capacity is determined by specific environmental variables (biodiversity, resource fluctuations, rainfall, land area, etc.) and by cultural variables (technology, cultural perception of resources, etc.). Williamson and Sabath go on to argue that long-term population stability depends on population size: a small population has a much higher probability of extinction than a large population. It therefore follows that islands with a high carrying capacity (i.e. those with a high potential population) are more likely to be permanently settled than those with a low carrying capacity. They are thus able to produce a mathematical model (Figure 4.1) relating carrying capacity, extinction probability and the likelihood of settlement: according to this model, there is a 'threshold population size' range between A (the population size above which islands will be settled) and B (the population size below which islands will not be settled). Between A and B is a range within which settlement may or may not occur. On the basis of empirical research in the Marshall Islands, Williamson and Sabath estimate values of A and B at seventy-eight and forty persons, respectively (i.e. islands with carrying capacities of less than forty people are unlikely to be colonised, though they may well be discovered and visited). As we have seen, however, the carrying capacity of an island depends on cultural as well as ecological variables,

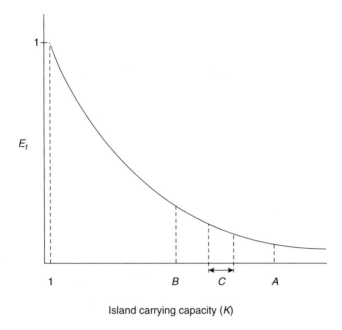

Figure 4.1   The generalised relationship between island carrying capacity (K) and probability of extinction (E) during the time interval t

Source: Williamson and Sabath 1984

Note: A is the population size above which populations always decide to settle, B is the population size below which populations always decide not to settle and C is the threshold region in which settlement decision passes from positive to negative

most significantly food procurement strategies. An island which could only support a population of thirty hunter–gatherers, for example, might support a population of seventy agriculturalists. The transition from hunting and gathering to farming, therefore, can be seen as a key threshold which makes island colonisation more likely. The 'secondary products revolution', which Sherratt (1979) places between the fourth and third millennia BC, could perhaps be seen as another. This may help to explain the pattern of colonisation of the Mediterranean islands noted in the last chapter.

The extent of contact and exchange between populations is a further factor which may influence the likelihood of island settlement (Williamson and Sabath 1984) since this may involve:

- Resource exchange when one population experiences resource depression.
- Refuge for individuals from resource depressed islands.
- Migration of individuals, mediated by sociological ties, when fertility in one population is threatened by small numbers in breeding age groups.

Williamson and Sabath's (1984) model depends largely on demography for, whilst carrying capacity determines the potential population of an island at a given moment in time, it is demography that determines the extinction probability of a community of a given size. Demography, however, is not (as Williamson and Sabath's model presupposes) an independent variable: it is affected by a wide range of cultural factors including marriage and incest rules and social attitudes to sexuality and child-rearing. As a result of this, island populations seem, in some cases, to rise at abnormally high rates following initial colonisation. This seems, for example, to have happened on Pitcairn Island following its colonisation by the Bounty mutineers in 1790 (Nicholson 1965), with population increasing at an average annual rate of 3.7 per cent between 1790 and 1810. Such cultural factors could have the effect of lowering the threshold population size (the values of A and B on Figure 4.1), permitting the settlement of islands with lower carrying capacities. If accelerated demographic growth continued for any length of time, however, it could lead to severe pressure on the resources of an island community, as seems to have happened on Pitcairn (see Chapter 2).

The effect of human communities on their natural environment is a subject of both academic and political concern, as highlighted by Bahn and Flenley (1992), in their book *Easter Island, Earth Island*, which uses resource depletion on Easter Island as a model for the depletion of the Earth's resources. The human effect on the environment is likely to be particularly marked in an island context. The concept of 'species equilibrium' (see Chapter 2) as outlined by MacArthur and Wilson (1967) implies that the establishment of a new species (including people) on an island is likely to be balanced by the extinction of species already present. Many of the Mediterranean islands had endemic faunas which had evolved in the absence of predators, and these are likely to have been particularly vulnerable to predation by human communities. Direct predation, however, is not necessarily the most significant factor in the extinction of island faunas. Diamond (1989) contrasts the 'blitzkrieg model' of species extinction (cf. Mosimann and Martin 1975), according to which extinctions occur rapidly through predation at a faster rate than reproduction of the victim species, with a 'sitzkrieg model', in which extinctions occur more gradually, as a result of more complex processes including habitat destruction and the introduction of non-native species, some of which compete with endemic species, and some of which act as vectors of disease. It will be argued below that the Mediterranean islands provide examples of both processes of extinction. Clearly, however, there is an important distinction to be drawn between hunter–gatherer and farming populations, since extinctions on the 'sitzkrieg model' are far more likely to occur in the context of agricultural activities, including large-scale clearance and the introduction of domesticated livestock. Following Diamond's model, therefore, we might expect to see a long period of coexistence between hunter–gatherer communities and some island endemics, followed by the rapid extinction of those species after the transition to food production.

Following the transition to food production, we might also expect to see a more dramatic transformation of the landscape of the Mediterranean islands as a result of human activities. Understanding of the effects of human communities on vegetation is hampered by the fact that much of the Mediterranean region is dominated by limestone soils, in which pollen is generally not well preserved (Vernet *et al.* 1984). Animal bone, on the other hand, is often well preserved in the alkaline conditions afforded by such soils, hence it is far easier to reconstruct the effect of human activities on island faunas than on island floras within the Mediterranean region. Evidence for human impact on the natural vegetation has been identified, for example in Corsica (Reille 1977), though this impact is not greatly different from that noted in comparable mainland contexts. In other island contexts, most notably in the Pacific, deforestation has resulted in large-scale erosion processes, which were in some cases detrimental, and in other cases benefi- cial, to the human communities concerned (Hughes 1985; Spriggs 1985; Kirch and Yen 1982). Whilst some authors (cf. Vermeule 1964; Hood 1970) have claimed to have identified evidence for similar processes on the Aegean Islands, this has been convincingly disputed by Bintliff (1977), who points out that present erosion rates are relatively insignificant, despite the fact that modern human activity is much more disruptive than at any other time.

This chapter will explore various aspects of the human ecology of the Mediterranean islands, and will attempt to ascertain the effect of human com- munities on these island ecosystems, and the extent of the 'island effect' on the relationships between human communities and their natural environment.

HUNTER–GATHERER COMMUNITIES AND MEDITERRANEAN ISLAND ECOSYSTEMS

As we have seen in the previous chapter, most of the Mediterranean islands seem not to have been permanently occupied by hunter–gatherer groups. This is not to say that the islands had not been discovered, or were not visited: the evidence from the Francthi Cave, for example, demonstrates clearly that the island of Melos was exploited as a source of obsidian in Early Mesolithic times (Perlès 1979), but there is no evidence for permanent settlement on this island before the Neolithic (Cherry 1981). Carrying capacity, therefore, seems to have been a greater barrier to island colonisation than maritime technology. The islands which have produced evidence for occupation by non-farming groups are Cyprus, Euboia, Halonissos, Skyros, Kephallenia, Corfu, Sicily, Elba, Sardinia, Corsica and Mallorca. Of these, we can exclude Euboia, Halonissos, Corfu, Sicily and Elba, since these would have been connected by land bridges to the mainland during the relevant periods. Kephallenia and Skyros, though significant in demonstrating Palaeolithic activity on true islands, have produced only limited archaeological data. The four remaining islands are significant in being amongst the largest islands in the Mediterranean.

Sardinia has been claimed as having the earliest evidence for settlement of a Mediterranean island. A series of flake tools from Anglona (Arca *et al.* 1982) have been identified as Lower Palaeolithic on typological grounds, but Cherry (1992) urges caution, since this typological identification is not backed up by any palaeontological or geostratigraphic evidence. Sondaar and Spoor (1982) have identified a sharp break in the Middle Pleistocene between the older Nesogoral fauna and the younger Tyrrhenicola faunal assemblage. The Nesogoral fauna is characterised by a number of species which became extinct in the Middle Pleistocene (*Nesogoral melonii, Rhagamys minor, Prolagus figaro* and *Macaca majori*), whilst the Tyrrhenicola fauna is characterised by a range of new species, including an endemic deer (*Megaceros cazioti*), a small canid (*Cynotherium sardus*), and the 'rabbit-rat' (*Prolagus sardus*). Sondaar and Spoor (*op. cit.*) point out that, unusually for Mediterranean island faunas, the later faunal assemblage includes a deer of mainland proportions, *Megaceros cazioti*. This suggests the existence of a large predator, possibly *Homo*, though it should be said that *Cynotherium sardus* is another candidate for this status (Sondaar *et al.* 1984). It is also suggested (Sondaar and Spoor *op. cit.*) that human activity could be responsible for the extinction of many of the species in the older, Nesogoral fauna, but this claim remains controversial (Vigne 1989; Cherry 1992), since there is no direct archaeological evidence for a human presence on Sardinia in the Middle Pleistocene.

The site of Corbeddu Cave (Sondaar *et al.* 1984) is of key significance in understanding the relationship between hunter–gatherer groups and the endemic fauna of Sardinia. Three distinct strata were identified in the excavation of this cave:

Level 1   35 cm of brown clay, with angular limestone pebbles, bones of domestic animals and some ceramic fragments. Charcoal from this deposit produced a radiocarbon date of: 6260 ± 180 bp (GRN-11433).

Level 2   50 cm of red clay with abundant remains of *Prolagus sardus*. Charcoal from this deposit produced a series of radiocarbon dates:
7860 ± 30 bp (UtC-301)
9120 ± 380 bp (GRN-11434)
11,040 ± 130 bp (UtC-250)

Level 3   40 cm of red clay with small quartz pebbles and fossils of *Megaceros cazioti*. Two radiocarbon dates were obtained from this horizon:
13,590 ± 140 bp (GRN-11435)
14,370 ± 190 bp (UtC-242)

There is no serious doubt that the assemblage from level 2 relates to human activity: the *Prolagus* remains from this horizon are associated both with artefacts and with human remains. The human remains consist of a right temporal bone and left maxilla from hall 2 of the cave, and a proximal ulna fragment from hall 1 (Sondaar and Spoor 1982; Klein-Hofmeijer *et al.* 1987). It is argued that these fossils have features which distinguish them from *Homo sapiens sapiens*:

> The articular eminence of the temporal bone is wide and flat in the mediolateral direction, whilst it is concave . . . in *Homo sapiens* . . . The maxilla is brachyranic and leptostaphylinic, an unusual combination . . . related to the proportions of the alveolar process. The molar alveoli are very large, both compared to the variation in *H. sapiens sapiens*, as well as compared to the alveoli of the anterior teeth.
>
> (Sondaar and Spoor 1982, 503)

It is argued that the morphology of these bones falls outside the known range of variation in *Homo sapiens*, and that these fossils therefore represent an endemic island species of *Homo*. This suggests a long period of isolation during the Pleistocene. This is an interesting, and by no means implausible suggestion, but caution clearly needs to be exercised (Vigne 1989; Cherry 1992) in making inferences of this nature on the basis of only three fossils, particularly since there is no independent evidence for an early Pleistocene human presence (Cherry 1992). The suggestion of endemic *Homo* on Sardinia would imply development in almost total isolation following the establishment of an island population: this would be compatible with an accidental colonisation, comparable to that envisaged for other animal species (cf. MacArthur and Wilson 1967).

According to Sondaar *et al.* (1984), the evidence from level 3 at Corbeddu provides some indication of a Pleistocene human presence, demonstrating the coexistence of humans with the Tyrrhenicola faunal assemblage: a number of deer bones from this deposit are considered to exhibit butchery marks (disputed by Cherry 1992), and the horizontal distribution of bones also suggests human influence. The age distribution of the deer remains from level 3 is anomalous, showing a low mortality rate among young animals, and peaks in mortality between the ages of 2 and 6, and between 9 and 12 years. Even if we ignore Cherry's (1992) very serious reservations about the anthropogenic nature of the material from level 3, this evidence does not significantly strengthen the arguments for a long period of isolated human development on Sardinia, since the radiocarbon determinations suggest that this activity dates to a relatively late stage in the Upper Palaeolithic, certainly much later than the claimed dates for the Anglona material.

Irrespective of this discussion concerning endemic *Homo*, and the possibility of Palaeolithic human settlement, the evidence from level 2 at Corbeddu suggests that hunter–gatherer groups were present on Sardinia from *c.* 1100 bp, and that they coexisted with endemic faunal species (notably *Prolagus sardus*) for at least 4000 years, into the Neolithic period. Whilst the ultimate extinction of this species can probably be attributed to the activities of

Neolithic communities, the long period of coexistence between *Homo* and *Prolagus* supports Diamond's (1989) 'sitzkrieg' model of extinction, rather than the 'blitzkrieg' model of Mosimann and Martin (1975). The extinction of *Prolagus* during the Sardinian Neolithic should probably be attributed to habitat destruction, and to competition from domestic animals, rather than to direct human hunting. Sondaar (1987) argues that the high reproductive rate of *Prolagus* (a relative of the rabbit) was a significant factor in permitting this extended coexistence of hunter–gatherer communities and endemic species, something which is rare on other Mediterranean islands.

The pattern on Corsica is broadly similar to that on Sardinia, unsurprisingly, since for much of the Pleistocene these two islands would have formed a single insular land mass. *Prolagus* is present on Corsica, as well as on Sardinia, and the evidence from the sites of Araguina-Sennola and Strette gives clear evidence for the exploitation of this species by hunter-gatherer groups (Vigne *et al.* 1981). The lowest level at Strette, which has produced *Prolagus* remains and a lithic assemblage (but no pottery) has given a radiocarbon date of 9140 ± 300 bp (LY-2837). The pattern of burning on the *Prolagus* remains from Araguina-Sennola shows clear evidence for human intervention, and suggests spit-roasting of whole animals (Vigne *et al.* 1981). As on Sardinia, there is clear evidence from both Strette and Araguina-Sennola for the survival of *Prolagus* into the Middle Neolithic.

The evidence from Cyprus is in marked contrast to that from Sardinia and Corsica. Until recently, the earliest evidence for human occupation on Cyprus dated to the Aceramic Neolithic. The presence of endemic mammal species on the island, including pygmy hippopotamus (*Phanourios minutus*) and elephant (*Elephas cypriotes*) had long been recognised, but these were considered to have become extinct prior to human contact. This picture has been radically changed by the discovery of the remains of these animals, in clear association with a lithic assemblage, at the rock shelter of Akrotiri Aetokremnos (Simmons 1991). The stratigraphy of this rock shelter is as follows:

Level 1   Sterile deposits deriving from roof collapse.

Level 2   A series of at least five microstratigraphic episodes. This deposit contains over 3000 hippo bones, as well as bird bones and marine shells, associated with a considerable quantity of chipped stone. This level has also provided a series of important radiocarbon dates:

9490 ± 120 bp (TX-5833A)
10,150 ± 130 bp (TX-5833B)
10,190 ± 230 bp (Beta-41405)
10,420 ± 85 bp (Beta-41000/ETH-7188)
10,480 ± 300 bp (Beta-41407)
10,485 ± 80 bp (Beta-41406/ETH-7160)

Level 3   Lenses of sterile sand.

Level 4   Midden, up to 50 cm in thickness, with large quantities
          of hippo and elephant bone, associated with a chipped
          stone industry. This level has also produced a series of
          radiocarbon dates:

          10,560 ± 90 bp (Beta-40382/ETH-7160)
          10,770 ± 160 bp (Beta-43176)
          11,200 ± 500 bp (UCL-194)

Simmons (*op. cit.*) stresses that the lithic assemblage from this site, dominated by thumbnail scrapers, burins and microliths, is quite unlike those found on sites of the Aceramic Neolithic period in Cyprus. The radiocarbon dates from Kalavassos-Tenta (the earliest dated site of the Aceramic Neolithic) suggest that this phase began at least a millennium after the final phase of activity at Akrotiri Aetokremnos (Todd 1987). The lithic assemblages suggest no continuity between the phase represented at Akrotiri and that at Kalavassos. On present evidence, therefore, it seems likely that the original human population either left Cyprus or became extinct, and that the island was reoccupied at the beginning of the Aceramic Neolithic period. The evidence from Kalavassos-Tenta also suggests that the pygmy hippopotamus and elephant had become extinct by this stage. It seems most likely, therefore, that these endemic species became extinct as a result of human hunting: an example of 'blitzkrieg' rather than 'sitzkrieg' in relation to Diamond's (1989) model.

On the Balearic island of Mallorca, remains of an extinct, endemic caprine, *Myotragus balearicus* have been identified. This species seems to have been present on the island for between 6 and 8 million years prior to the arrival of human communities, and Juniper (1984) suggests that the presence of *Myotragus*, in the absence of signficant predators, would have had a marked effect on the natural vegetation of the island, with tree cover being effectively confined to inaccessible slopes, and the vegetation cover being dominated by heavily armed spiny and poisonous plants, such as the endemic *Teucrium subspinosum, Anthyllis fulgurosis, Astragalus balearicus, Sonthus spinosus, Centaurea balearica* and *Genista lucida*. Human communities arrived on Mallorca at a relatively late stage. Waldren (1982) has identified an 'early settlement phase' on the basis of the evidence from Son Muleta and Son Matge. This phase is apparently aceramic, and is characterised by the absence of domesticated animals. The cave of Son Matge has produced the most significant evidence, including a bed of *Myotragus* coprolites. This deposit is up to 125 cm in depth, and has abrupt edges, suggesting that the animals were deliberately corralled. Butchered bones were associated with the deposit, suggesting that animals were occasionally slaughtered in the cave, and the existence of a number of artificially trimmed horns provides further evidence for deliberate management of the herds. The *Myotragus* coprolite bed has provided two radiocarbon dates:

6680 ± 120 bp (QL-29)
5820 ± 360 bp (SCIC-176)

Butchered *Myotragus* remains were also found in kitchen midden deposits overlying the coprolite beds at Son Matge. These deposits have produced the following radiocarbon dates:

5750 ± 115 bp (I-5516)
4650 ± 120 bp (QL-988)

Whilst the deposits of the 'early settlement phase' have produced neither pottery nor the remains of domestic animals, the radiocarbon dates from Son Matge suggest a chronological overlap with Neolithic cultures in the western Mediterranean, and the evidence for corralling suggests that these settlers were at least familiar with techniques of animal husbandry. Clutton-Brock (1984) goes so far as to suggest that the abundance of *Myotragus* may account for the fact that the earliest settlers did not introduce domestic animals.

The Canet Cave has produced possible evidence for human exploitation of *Myotragus* at an earlier stage (Kopper 1982). The earliest dated cultural layer on this site provided a radiocarbon date of 9170 ± 570 bp (P-2408), whilst a higher cultural level produced a date of 6370 ± 320 bp (Beta-6948). In both cases, however, the high standard deviations render the dates of questionable validity. The evidence from the Canet Cave does raise the possibility, however, that the corralling of *Myotragus* at Son Matge may not represent the earliest human exploitation of this species.

The most recent evidence for the existence of *Myotragus* is from Son Matge, in a deposit which has provided a radiocarbon date of: 4093 ± 398 bp = 3300–2040 cal. BC (BM-1408). Even if we ignore the radiocarbon dates from Canet, this must represent a coexistence of *Myotragus* with human communities of at least 2800 years and, as with *Prolagus* on Sardinia/Corsica, this species probably became extinct as a result of habitat destruction and competition from introduced animals, rather than through direct human exploitation.

These three case studies, on Sardinia/Corsica, Cyprus and Mallorca, present some interesting contrasts concerning the relationships between hunter–gatherer communities and island endemics. Given the specialised nature of island faunas, and the fact that, in all three cases, endemic herbivorous mammals had evolved in the absence of predators, one might expect the impact of human contact to have been catastrophic for the species concerned. In fact, the archaeological evidence suggests that this was the case only on Cyprus where, we must imagine, pygmy hippo and elephant were highly vulnerable to human hunting, and were unable to reproduce fast enough to replace the individuals killed by human hunters. On Sardinia/Corsica, *Prolagus* was perhaps spared this fate on account of its high reproductive rate, and the evidence suggests a long period of coexistence with hunter–gatherer communities. The situation of *Myotragus* on Mallorca is a little different,

71

since the evidence from Son Matge suggests deliberate husbandry rather than simply hunting. Both on Sardinia/Corsica and on Mallorca, it seems that the impact of hunter–gatherer communities on island ecosystems was minimal, certainly far less catastrophic than the impact of later farming communities. The example of Cyprus shows, however, that this impact depends very much on the biology of the individual species concerned, and that the concept of 'species equilibrium', as defined in the island biogeography model, is too simplistic to adequately explain the complex ecological relationships between human communities and island faunas.

THE TRANSITION TO FOOD PRODUCTION

Between 7000 and 5000 cal. BC, human communities in the Mediterranean region adopted a variety of food-producing strategies, which included the cultivation of cereals (and probably legumes, though these are less archaeologically visible) and the rearing of domestic animals. The transition to food-production had a number of effects on island communities. First, it seems to have resulted in the colonisation of many more islands (see Chapter 3). This is probably a function of carrying capacity (cf. Williamson and Sabath 1984), since food-production (probably combined with exploitation of natural, especially marine, resources) would allow a small island to support a greater number of people. Second, it seems to have placed considerable stress on island ecosystems, since food-production strategies represent a far more active intervention in the environment than hunting and gathering strategies. This is most dramatically seen on Sardinia/Corsica and on Mallorca, with the extinction of *Prolagus* and *Myotragus*, which had previously coexisted with human communities. Clearance of the natural vegetation, competition for grazing between native and introduced species, diseases introduced by domestic livestock, and perhaps direct predation by introduced animals such as dogs, are likely to have had a combined effect on these endemic species. Farming communities, unlike hunter–gatherers, are likely to have had a marked effect on the flora, as well as on the fauna, of the Mediterranean islands, transforming the basis of the natural food chain.

The adoption of food-producing strategies in an insular context involves processes which do not apply to mainland communities. Domestic livestock, in particular, have to be transported over water, requiring boats of a certain size and strength and, if the voyage is more than a day in length, a means of feeding and watering livestock on board. We might expect, therefore, to identify contrasts between the process of transition from food procurement to food-production strategies on the Mediterranean islands and on adjacent mainland areas. Since even the most sophisticated food-production strategies have their basis in the natural environment, we might also expect that the limited biodiversity, particularly of the smaller Mediterranean islands, would have influenced the type of food-producing strategies adopted.

Table 4.1 Hunted species as a component of meat diet in Neolithic Corsica

Phase	% of meat acquired by hunting	Species represented (%)	
Pre-Neolithic	100[1]	Monk seal	70
(L18 at Araguina-Sennola)		*Prolagus*	30
Cardial	2	*Prolagus*	
(L7 at Basi)		Fox	
		Goose	
(L14 at Strette)	4	*Prolagus*	
		Marine molluscs	
Epi-Cardial	1	*Prolagus*	
(L13 at Strette)		Oyster	
(L17 at Araguina-Sennola)	28	*Prolagus*	27
		Sea birds	0.7
		Land birds	0.42
		Rodents	0.01
Middle Neolithic	14	*Prolagus*	0.6
(L16-14 at Araguina-Sennola)		Other land species	0.1
		Marine species	13.3
(L23-20 at Scaffa-Piana)	2	*Prolagus*	
		Birds	
		Fish	
		Marine molluscs	

Source: Vigne 1987

Note: 1 A small number of sheep and pig bones have in fact been recovered from the upper part of this horizon, separated by a rock fall from any risk of contamination (Lewthwaite 1990)

Table 4.2 Domesticated species as a component of meat diet in Neolithic Corsica

Phase	% of meat provided by domestic animals	Species represented (%)	
Cardial			
(L7 at Basi)	99	Sheep/goat	49.5
		Pig	49.5
Epi-Cardial			
(L17 at Arraguina-Sennola)	72	Pig	57.6
		Sheep/goat	14.4
Middle Neolithic			
(L16-14 at Araguina-Sennola)	86	Cattle	51.6
		Sheep/goat	10.0
		Pig	24.1

Source: Vigne 1987

In examining the transition from food procurement to food production, Sardinia and Corsica are perhaps good starting points, since we have evidence for continuity of occupation spanning the transition period, with a series of important stratified sites, most notably Corbeddu, Strette and Araguina-Sennola. Vigne (1987) has carried out a detailed analysis of the faunal remains from the stratified sites of Araguina-Sennola, Basi, Strette and Scaffa Piana on Corsica. The results of this analysis are shown on Tables 4.1 and 4.2.

This evidence suggests that, whilst domesticated pig and ovicaprids were introduced to Corsica during the Cardial phase, domestic cattle were not introduced until the Middle Neolithic. The evidence from Sardinia is broadly comparable (Sanges 1987), though red deer were apparently introduced, alongside domesticated animals. Lewthwaite (1990) discusses the 'filter effect' of the Mediterranean islands, pointing out that domesticated cattle do not appear in southern France until the Middle Neolithic, and suggesting that the islands of Corsica and Sardinia played a crucial role in the transmission of elements of the Neolithic package from east to west. Lewthwaite (1985a) also stresses the lack of evidence for cereal production and village settlement in Corsica prior to the Middle Neolithic period. The evidence therefore suggests a second significant transformation in the Middle Neolithic of Corsica, and it is clear from the faunal assemblages that *Prolagus* did not become extinct until after this transition. It is conceivable that the 'secondary products revolution' was an element of this transformation. Vigne (1987) points out that, in the Middle Neolithic levels at Araguina-Sennola, the age structure of the sheep/goat population suggests the rearing of these animals for milk production. As regards the Early Neolithic (Cardial and Epi-Cardial), Lewthwaite (1985a) suggests a model according to which indigenous Corsican communities, with an economy based on the exploitation of *Prolagus* and marine resources, adopted innovations (including the domestication of sheep and goat) which were capable of increasing and diversifying their resource base. The true 'commitment to agriculture' occurred at a later stage. The faunal evidence outlined by Vigne (1987), whilst offering broad support to many aspects of Lewthwaite's model, suggests a somewhat more complex set of processes. The dramatic decline in the significance of hunted resources, from 100 per cent in the pre-Neolithic horizon at Araguina-Sennola, to 2–4 per cent in the Cardial horizons at Basi and Strette suggests a radical change in economic strategy, rather than a simple diversification of the resource base. During the Epi-Cardial and Middle Neolithic phases on Corsica, it is only at Araguina-Sennola that we see evidence for hunting as a significant element of the economy. Vigne (1987) draws attention to the anomalous nature of this assemblage, and suggests that Araguina served as a late-winter and spring hunting camp: the cultural layers are separated by sterile lenses, suggesting intermittent occupation, and the assemblage includes remains of migratory birds, which are most likely to have been present in late-winter. This Early Neolithic transformation must surely be seen as evidence for a major economic intensification, in which hunting, previously an all-year activity and the mainstay of the economy, became a seasonal and supplementary strategy. Given the long time period covered by the Pre-Neolithic in Sardinia and Corsica, and the apparently stable nature of both human communities and natural resources during this period, it seems hardly likely that this sudden transformation was forced on human communities by the limitations of the island ecosystem. It would perhaps be more plausible to suggest that intensification was related to increasing social complexity, and that new strategies were adopted in order to meet increasing social demands for surplus production (cf. Bender 1978). The Middle Neolithic transformation, which involved the adoption of cereal cultivation, the introduction of domestic cattle, and perhaps the exploitation of secondary products (particularly dairy products) can be seen as a further intensification. It is perhaps no coincidence that this second phase of intensification is followed by the appearance of megalithic monuments on Corsica (Lewthwaite 1985a). Whilst similar patterns of intensification can be identified in mainland contexts, it is likely that established island populations, such as those on Corsica and Sardinia, would adapt food-producing strategies borrowed from elsewhere to suit the resources and constraints of the island ecosystem. Lewthwaite (1985a) draws attention to the the faunal assemblage from the Middle Neolithic settlement of Terrina IV, on Corsica (Camps 1981), pointing out that the faunal proportions differ strikingly from the expected cereal/ovicaprid regime. Ovicaprids represent less than 25 per cent of the sample, with cattle and pig predominant. Lewthwaite therefore argues for a 'forest Neolithic' on Corsica, suggesting that cattle and pig are suited to the forest edge and marshlands, rather than to stubble field and mountain grazing, and pointing also to equipment which may have been used for processing acorn flour.

The Balearic Islands are of interest since here, as on Corsica and Sardinia, we see evidence for continuity between 'pre-Neolithic' and Neolithic communities. The situation on the Balearics, however, differs from that on Corsica and Sardinia, in that these islands were settled considerably later, and by people who seem to have had at least some understanding of animal husbandry. Thus at Son Matge, in a context dating to the seventh millennium bp, we see evidence (Waldren 1982) for the corralling and deliberate management of herds of *Myotragus*. There is no evidence, however, from this 'Early Settlement Phase', for the introduction of domestic animals, and no direct evidence for cereal cultivation (cereal pollen is present in pollen samples from the early settlement phase at Son Muleta, but this could conceivably have been blown in from the Iberian mainland). If we accept the radiocarbon dates from Son Matge and Son Muleta (see above) as representing the earliest human settlement of the islands (and we cannot rule out the possibility of an earlier, and as yet undiscovered, cultural horizon), then any founder population from the Iberian mainland is likely to have had some knowledge of both cereal cultivation and animal husbandry. We must then ask why domesticated animals were not introduced at this early stage, and it may simply be that the abundance of *Myotragus* made this unnecessary. The islands may have been discovered as a result of fishing expeditions, and may have been settled precisely because of the abundance of an endemic species which, having presumably little fear of humans, could easily be herded and corralled.

The earliest evidence for introduced domestic animals on the Balearic Islands is from the 'Neolithic Early Ceramic phase' (Waldren 1982) at Son Matge. This horizon has produced two radiocarbon dates:

4650 ± 120 bp = 3590–3160 cal. BC (QL-988)
4093 ± 392 bp = 3300–2040 cal. BC (BM-1408)

Within these hearth deposits at Son Matge were found bones of domesticated goat, pig, cattle and *Myotragus*. The associated pottery has been compared by Waldren (1982) to material from Middle/Late Neolithic sites in southeastern Iberia, such as Tabernus, Abrigo de Ambrosio (Almeria) and Cueva de l'Or (Alicante). As we have already seen, the evidence suggests the extinction of *Myotragus* shortly after the beginning of the Neolithic Early Ceramic Phase (the latest dated remains of *Myotragus* being from the horizon which also produced the radiocarbon date BM-1408). The Early Ceramic phase also sees the appearance of open-air settlements, as at Ferrandell-Oleza (Waldren 1982). The faunal assemblage from this site (Clutton-Brock 1984) is dominated by ovicaprids, but bones of pig, cattle and dog are also present: no wild species are represented. Whilst Clutton-Brock (1984) suggests an increasing reliance on domestic animals as *Myotragus* was hunted to extinction, the long period of coexistence between *Myotragus* and human communities, and the absence of any evidence for a gradual introduction of livestock, suggest that the decline of *Myotragus* is more likely to have been a consequence, rather than a cause, of the introduction of domestic livestock. As on Corsica and Sardinia, the evidence suggests that island communities initially adapted mainland strategies to the local environment (animal husbandry techniques applied to *Myotragus*, for example), but that later intensification, perhaps for social reasons, led to the adoption of the full Neolithic package, to the detriment of endemic species. The human impact on the fauna of the Balearic Islands, however, did not end with the Neolithic, and the introduction of domestic livestock. Sanders and Reumer (1984) list a total of eleven 'wild' species present on the island of Menorca today, which seem not to be native to the island. The faunal remains from the site of Rafal Rubi, dating to the mid-second millennium cal. BC, includes four such species: *Eliomys quercinus* (the dormouse), *Mus musculus* (the common mouse), *Apodemus sylvaticus* (the field mouse) and *Oryctolagus cuniculus* (the rabbit), as well as the endemic frog, *Baleaphryne talaioticus*, and a lizard, *Lacerta (Podaris)* sp. The same species are present in the assemblage from Binicalaf (dating to the third century cal. BC), and from the lowest level at Taula Torralba d'en Salort, dating to the early first millennium cal. BC (the absence of the rabbit, *Oryctalagus*, from the latter site is explained by the fact that this assemblage is derived entirely from owl pellets). The second century BC was marked by the Roman conquest of the Balearic Islands, under Quintus Caecilius Metellus (123–122 BC). In subsequent levels at Taula Torralba d'en Salort (Levels I–IV), the faunal assemblages include *Rattus rattus* and

Crocidura suaveolus (the white-toothed shrew), species which were presumably introduced by the Romans. Other species seem to have been introduced at a later stage, notably *Mustela nivalis* (the weasel) and the gecko, *Hemidactylus* sp. The phenomenon of 'species equilibrium' (cf. MacArthur and Wilson 1967) can perhaps be seen in the sharp decline in the relative proportions of *Eliomys* between levels IVa (47.1 per cent) and IVb (5 per cent) at Taula Torralba d'en Salort.

Lewthwaite (1990) draws a distinction between the transition to food procurement on the islands of the west Mediterranean and on the islands of the east Mediterranean (including the Aegean). In essence this is a distinction between islands (such as Corsica, Sardinia and the Balearics) which already supported human populations and those (such as Crete, and most of the Aegean Islands) which did not. Alone in the eastern Mediterranean, Cyprus does seem to have been settled by hunter–gatherer communities but, as we have seen, there is no evidence for continuity between these groups and the Early Neolithic. On islands which were already settled, local insular communities were able to adapt and modify food procurement strategies to their own specific ecological and cultural needs, whereas on those that were being settled for the first time (or, as on Cyprus, following an apparent phase of abandonment), the Neolithic package arrived in a 'ready-made' form.

This second process is perhaps most clearly seen on Crete (Broodbank and Strasser 1991). Agriculture was established on Crete in the late eighth and early seventh millennia cal. BC, and represents one of the earliest successful maritime transfers of a full farming economy. There is no convincing evidence for pre-Neolithic human settlement on Crete, though as always negative evidence is not conclusive. The earliest Neolithic evidence from Crete is from stratum X at Knossos (Evans 1964; 1968; 1971b). This layer is aceramic, but has abundant remains of ovicaprines, cattle and pig, as well as carbonised grains of bread wheat (*Triticum aestivum*), all of which are exogenous to the Aegean. Broodbank and Strasser (*op. cit.*) argue that the first domesticates at Knossos represent 'the full Anatolian/Balkan Neolithic faunal and floral package, without any indication of filtering' and stress the similarities between the faunal assemblage at Knossos and those from contemporary mainland sites, such as Achilleon I, Argissa Magoula, Nea Nikomedia and Franchthi. Given the position of Crete (Figure 1.2) in relation to the Greek and Anatolian mainlands, one might expect this colonisation to have taken place using stepping-stone islands, either from Anatolia via Rhodes, Karpathos and Kasos, or from Greece via Kythira and Andikythira. Surprisingly, however, the material from stratum X at Knossos predates the earliest known Neolithic material from these islands, leading Broodbank and Strasser (*op. cit.*) to postulate that Crete (probably already known through fishing expeditions) was deliberately targeted for colonisation on account of its size: the smaller islands may have provided useful stopping-off points, but were considered too small to settle. Crete is relatively distant from the mainland (100 km from Cape Malea in the Peloponese, 185 km from Knidos in Anatolia) but, because of the existence of stepping-stone islands, the greatest single stretch of water to be crossed is 48 km. Broodbank and Strasser (*op. cit.*) estimate an initial founder population of around forty people, based on the size of the Knossos settlement itself (cf. Evans 1971b, Broodbank 1992). A founding group would need to carry around 250 kg of grain per person, and would need to take a minimum of 5 to 10 pairs of each animal species in order to preserve enough genetic diversity to give rise to a viable population (Foose 1983). This implies that the founding population which colonised Crete would have to carry around 10,000 kg of grain, 10 to 20 pigs, 10 to 20 ovicaprines and 10 to 20 cattle, probably stopping on smaller islands in order to feed and water the livestock.

The Neolithic reoccupation of Cyprus seems to have involved a similar introduction of an integrated Neolithic package. The earliest dated Neolithic site on the island is Kalavassos-Tenta (Todd 1987), where the radiocarbon dates (see Chapter 3) suggest an initial occupation at *c.* 7500 cal. BC. The faunal assemblages from this site, and from the nearby site of Khirokitia (Le Brun 1984) include domestic sheep, goats and pigs, as well as foxes and fallow deer, which also seem to have been introduced. Cattle are absent, but are in any case not ideally suited to the arid environment of Cyprus.

The Cycladic island of Antiparos, together with the off-shore islet of Saliagos, seems, like Crete, to have been occupied for the first time in the Neolithic (Evans and Renfrew 1968). The earliest horizon on this site produced two radiocarbon dates:

6257 ± 81 bp = 5280–5080 cal. BC (P-1396)
5949 ± 87 bp = 4900–4758 cal. BC (P-1333)

As on Crete and Cyprus, the full range of domestic animals are present in the earliest Neolithic levels, suggesting the importation of the full Neolithic package to the island. The overall composition of the mammalian faunal assemblage is shown on Table 4.3.

No evidence for hunting was established, but sieving of deposits from pit A, and from square N3, showed that fishing was a vital part of the economy, as shown by Tables 4.4 and 4.5.

Although, in faunal terms, these assemblages are fully 'Neolithic' (in that the full range of domestic animals are present), the assemblage from pit A in particular suggests that stock raising was of minor importance compared to fishing, and this is a point to which we shall return.

It should come as no surprise to find that the human impact on Mediterranean island environments increased markedly following the transition to food production. The extinction of *Prolagus sardus* on Sardinia and *Myotragus balearicus* on Mallorca has already been mentioned, and can probably be related, directly or indirectly, to the introduction of domestic animals. Because of the predominantly alkaline limestone soils of most of the Mediterranean islands, bones are generally well preserved, whereas pollen

Table 4.3 Composition of the mammalian faunal assemblage from the Neolithic deposits
at Saliagos

Species	Animals (%)	% of Meat diet
Sheep/goat	83.5	65
Cattle	3.5	20
Pig	12.1	15

Source: Evans and Renfrew 1968

Table 4.4 Faunal assemblage from pit A at Saliagos

Species	Minimum no. of individuals	% Meat diet
Sheep/goat	17	6.6
Cattle	1	2.8
Pig	4	2.4
Tunny	48	88.0
Patella	1900	0.09
Monodonta	340	0.005
Murex	140	0.0007

Source: Evans and Renfrew 1968

Table 4.5 Faunal assemblage from square N3 at Saliagos

Species	Minimum no. of individuals	% Meat diet
Sheep/goat	22	19.3
Cattle	6	38.2
Pig	10	13.6
Tunny	7	28.5
Patella	1970	0.21
Monodonta	650	0.02
Murex	216	0.55
Cerastoderma	53	0.002

Source: Evans and Renfrew 1968

evidence is limited, so that we know far more about the human impact on the fauna than on the flora. There has, as a result, been a certain degree of speculation concerning the possible impact of human communities (and especially farming communities) on the vegetation of the Mediterranean islands. Most of the Mediterranean islands (particularly those of the eastern Mediterranean) are today largely barren of trees, and some authorities have seen this as a result of human activities (cf. Vermeule 1964; Hood 1970). If large-scale deforestation did occur in prehistoric times, one might expect this to have given rise to rapid erosion and landscape transformation, as was clearly the case, for example, on some of the Pacific islands (cf. Kirch and Yen 1982; Spriggs 1985). Bintliff (1977) has criticised the 'Garden of Eden' model for the Aegean islands developed by Vermeule (1964), according to which the islands were originally covered by lush forest: he argues that the absence of trees on most of the Aegean islands can be explained by environmental, rather than anthropogenic factors: 'the characteristic barren and thin-soiled appearance of South-East Greece and the Aegean Islands is largely a product of the Mediterranean climate, and owes little to the hand of Man' (Bintliff 1977, 50). According to Bintliff's model, the absence of trees is largely a factor of poor soil development, owing to extreme dryness. Limited winter rainfall carries some nutrients down into the soil, but this water, with its mineral content, is drawn back up to the surface during the dry summers, so that, in pedological terms, an AC profile develops, rather than an ABC profile, which could support a greater range of vegetation (Anastassiades 1949). The loss of stored water also means that physical weathering of the parent rock is more significant than chemical weathering, so that rock is broken up and carried away as small particles unsuitable for soil formation.

The situation in the west Mediterranean islands is rather different. On the Balearic Islands, for example, Juniper (1984, 148) states that: 'In the absence of herbivorous large mammals, and given the relatively high rainfall and rich limestone rock formations of Mallorca, we would have expected a virtually complete forest canopy over the island'. He argues, however, that browsing by *Myotragus* is likely to have destroyed the forest cover, in all but the most inaccessible slopes of the island, prior to the arrival of human settlers. Again, therefore, deforestation cannot be attributed to human activity.

On Corsica and Sardinia, by contrast, palynological evidence does suggest that primary forest was cleared by human communities. Reille's (1977) palynological research at the Creno Lake on Corsica suggests that a natural cover of *Quercus* woodland started to decline dramatically at around 2500 bp (i.e. 700–800 cal. BC). The decline of *Quercus* is followed by an increase in *Betula*, later partially replaced by *Alnus glutinosa*, a cycle which Reille attributes to clearance by fire, since *Betula* is very tolerant of burnt soils, and develops in cleared areas but, being intolerant of shadow, tends at a later stage to be replaced by other species such as *Alnus*. This sequence is followed by a recovery of *Quercus*, but a second *Betula/Alnus* cycle is observed in the pollen record at around 1600 bp (i.e. 400–500 cal. AD). What is surprising about these results is not that the observed changes occured, but that they did not occur at an earlier stage. It is possible, therefore, that the 'forest Neolithic' strategy postulated by Lewthwaite (1985a) was less destructive than the more normal Mediterranean cereal/ovicaprid regime.

The archaeological evidence suggests an expansion of the pastoral sector on the islands of the west Mediterranean during the third and fourth millennia cal. BC, and this has been seen (cf. Lilliu 1975) as an ecological adaption to (primarily upland) areas, marginal to agriculture. This expansion is particularly evident on Sardinia and Corsica (Lewthwaite 1984a), where the transition from the Ozieri culture to the Nuraghic culture is marked by increasing evidence for occupation of upland plateaux. Whilst little detailed palaeoeconomic research has been undertaken, Webster and Michels (1986)

Table 4.6 Faunal assemblage from Iron Age horizon (900–500 cal. BC) at Nuraghe Toscono

Species	Minimum no. of individuals	% Meat Diet
Sheep/goat	50	23
Cattle	19	55
Pig	6	8
Red deer	6	8
Roe deer	19	6

Source: Webster and Michels 1986

have carried out an important study of the faunal remains from the (relatively late) nuraghe of Toscono. The composition of this assemblage is shown on Table 4.6.

This assemblage can be profitably compared with that from the Middle Neolithic horizon at Araguina-Sennola on Corsica (Vigne 1987 and see Table 4.2), in which pig represents 24.1 per cent of the meat diet and ovicaprids only 10 per cent, and with that from the Middle Neolithic site of Terrina IV on Corsica (Camps 1981), where again, cattle and pig predominate. Webster and Michels (*op. cit.*) also note that both caprines and cattle in the Toscono assemblage had been slaughtered as mature animals, suggesting secondary product exploitation. Similar stock ratios and mortality patterns have been noted at Tharros and Palmavera (Fedele 1980). The palaeobotanical evidence from the Toscono middens demonstrates the presence of barley and wheat, as well as wild, and possibly domesticated, legumes, on the basis of which Webster and Michels (*op. cit.*) suggest a strategy of small-scale pastoral transhumance, linked into a sedentary farm community. Pastoralism, therefore, seems only to be one element of the upland economy of Nuraghic Sardinia, albeit, as Lewthwaite (1984a) stresses, the most archaeologically visible element. Lewthwaite (*op. cit.*) sees the introduction of the ard (an element of the secondary products revolution) in the mid-fourth millennium cal. BC as a possible factor enabling communities to disperse and colonise areas below the threshold of effective hoe cultivation. He goes on to argue that reproduction and possibly day-to-day management of livestock, particularly plough oxen, are likely to have been controlled by an elite, since it would be impossible for every family to maintain and reproduce such beasts. This control over a significant element of the means of production may have been a factor in the emergence of assymetrical social relations in the uplands of Sardinia and Corsica.

THE ROLE OF COASTAL RESOURCES

So far, the discussion in this chapter has concentrated on terrestrial food resources, whether procured through hunting and gathering or by cultivation and stock rearing. Islands, however, are by definition maritime environments, and one might therefore expect that the resources of the sea would be at least as important to island communities as the resources of the land. Indeed this is true of many island communities in the Mediterranean even today. The exploitation of coastal resources may offset the effects of reduced biodiversity and resource limitation which MacArthur and Wilson (1967) see as a fundamental characteristic of island ecosystems. Most of the Mediterranean islands are small (see Figures 1.1 and 1.2), giving all communities potentially direct access to the sea, and on the larger islands, such as Sardinia and Corsica, the earliest settlements are often in coastal areas. Contrary to MacArthur's and Wilson's (1967) model, the resources available to such communities may in fact be considerably more abundant and more diverse than those available to contemporary communities in land-locked continental areas. Even in situations where fishing seems not to have been an important part of the economy (as for example, on Early Neolithic sites such as Kalavassos-Tenta and Khirokitia on Cyprus), the proximity of the sea could provide an important nutritional 'safety net' in the event of crop failure or the decimation of domestic animals by disease or drought. The exploitation of coastal resources may also have been a significant factor in the discovery of the more remote Mediterranean islands: the evidence from the Francthi Cave (Perlès 1979) demonstrates that deep-sea fishing was an important economic activity, at least in the Aegean, from a relatively early stage within the Mesolithic, and fishing expeditions may well have resulted in the discovery of islands such as Crete and Malta. It is likely that the role of coastal resources in prehistoric economies in the Mediterranean (as elsewhere) has been underestimated in the published literature: mammal bones are large and visible, and are unlikely to go unnoticed, whereas fish bones, on the other hand, are small, inconspicuous, and are only likely to be recovered by sieving. The scale of this problem is evident from the results of the excavation at Saliagos (Evans and Renfrew 1968): for logistical reasons, it was not possible to sieve the soil from the entire site, and the overall composition of the faunal assemblage suggests (Table 4.3) that sheep and goat represent 65 per cent of the meat consumed on the site. This proportion is comparable to that for many contemporary sites on the Greek and Anatolian mainland. When we look, however, at the faunal assemblages from the two deposits which were sieved (Tables 4.4 and 4.5), we find that tunnyfish accounts for 88 per cent of the meat in pit A, and 28.5 per cent in square N3, with sheep/goat accounting for 6.6 and 19.3 per cent, respectively. This gives a very different picture of the economy of the site, and it is unfortunate that comparable data are not available for many Mediterranean islands.

The clearest evidence for the exploitation of coastal resources is from the Aegean, and the site of Saliagos is of particular interest in this respect. Saliagos is a small islet in the channel between the islands of Paros and Antiparos. Since present sea levels are likely to be around 6 m higher than those in the Neolithic, it is likely that Saliagos originally formed part of an isthmus linking

Paros and Antiparos. Excavations by Evans and Renfrew (1968) revealed evidence for settlement with three distinct strata. The lowest stratum contained hearth areas and ephemeral remains of buildings, along with two refuse pits. Finds include painted pottery, obsidian tools, stone axes, and stone and bone beads. Charcoal from this deposit produced two radiocarbon dates:

6257 ± 81 bp = 5300–5100 cal. BC (P-1396)
5949 ± 87 bp = 4900–4770 cal. BC (P-1333)

The second stratum contained more substantial structural remains, including a circular structure and a rectangular house. The most significant structural remains, however, were in the uppermost level, where a rectangular enclosure was found (255 m² in area), with a substantial circular 'bastion' or tower. The finds from this level include a marble figurine (the 'Fat Lady of Saliagos'), a spondylus bracelet and four axes.

The significance of fishing in the economy of the Saliagos settlement is clear from the faunal assemblages which have already been outlined (Tables 4.4 and 4.5). It has also been suggested that many of the artefacts from the site reflect specialised fishing activities: a number of stone weights were interpreted as net-sinkers, whilst the obsidian assemblage is dominated by 'Saliagos points', which could be interpreted as leister prongs for spearing fish. The characteristic nature of the obsidian assemblage from Saliagos allowed Evans and Renfrew (*op. cit.*) to define the 'Saliagos culture', with similar assemblages elsewhere in the Cycladic archipelago, for example at Mavrispilia on Mykonos, Vouni on Antiparos and Agrilia on Melos. In each of these assemblages, the 'Saliagos point' is the most characteristic element. Bintliff (1977) suggests that the Saliagos culture as a whole reflects the activity of fishing communities following shoals of migratory fish, particularly tunny. He points out that the major fish catches in the Mediterranean today are made with migratory species such as tunny, sardine and mackerel which follow constant routes, allowing fishing communities to plan large-scale expeditions. Classical writers such as Aelian and Oppian refer to tunny towers, used for observing the movement of the shoals, and Bintliff (1977) suggests this as an interpretation of the circular tower at Saliagos. It is suggested that the widespread distribution of Melian obsidian in the Aegean from an early stage may reflect fishing territories rather than trade networks as such: Bintliff (*op. cit.*) suggests that the same communities who fished at Saliagos and Mykonos may have visited Melos and extracted obsidian. These fishing networks may also have played an important role in island colonisation, both through the discovery of new islands in the course of fishing expeditions and through the economic importance of fishing. Bintliff points out that one of the problems faced by a farming community attempting to colonise an island is that of moving within a single season, whilst maintaining a secure economic base: the seed grain which the colonists took to the island would need time to grow, and the livestock would probably be recently weaned young animals which would need to mature before becoming productive and forming a breeding pool. One answer to this problem might be to rely on marine resources. The communities of the Saliagos culture, however, were by no means exclusively reliant on coastal resources. On the contrary, the evidence from Saliagos itself demonstrates that they had the full range of domestic animals and cereal crops. This combination of abundant coastal resources and the normal range of domesticates may have allowed communities on the Aegean islands and littoral to produce a substantial surplus, which may well have provided the basis for the development of urban centres such as Phylakopi on Melos in the Early Cycladic period.

It is unlikely that the Saliagos culture, with its reliance on migratory fish shoals, was unique in the prehistory of the Mediterranean. The Dalmatian coast is an important area for tunny fishing today (Morgan 1956), as are the Aeolian Islands and the Sicilian coast (Klemmer 1959). Unfortunately, however, owing to the lack of direct archaeological evidence at this stage, we can only speculate on the possible importance of marine resources in these areas.

SYNTHESIS: INSULARITY AND HUMAN ECOLOGY IN MEDITERRANEAN PREHISTORY

Having looked in detail at some of the evidence for the ecological basis of prehistoric island communities in the Mediterranean, it is necessary to return to the more general question of the effect of insularity on human populations. Was the ecology of Mediterranean island communities significantly different from that of populations in mainland areas adjoining the Mediterranean Sea? If so, what were the specific effects of insularity on the ecology of these communities? On the basis of the general discussions in Chapters 1 and 2, it is possible to make a series of predictions concerning the likely effects of insularity, which can then be assessed in relation to the archaeological evidence outlined in this chapter.

1 *Following MacArthur's and Wilson's (1967) model, we would expect human communities on islands to demonstrate adaptations to an environment characterised by reduced biodiversity.*

According to the 'theory of island biogeography', island environments, particularly the environments of small, remote islands, should be characterised by reduced biodiversity, to which human communities, like other animal populations, would have to adapt. In fact, however, as we have seen, this depends very much on the extent to which coastal resources (the diversity of which will not be affected by insularity) were exploited. The Neolithic island communities of the 'Saliagos culture', for example, seem to have flourished in an environment of plenty, certainly very much richer than the environment of contemporary groups in inland areas of Anatolia or Greece.

It is also true to say that, whilst many of the Mediterranean islands were characterised, prior to human settlement, by reduced biodiversity, the majority were colonised by farming communities who, to a large extent, imported their own ready-made ecosystems, as on Crete, where the full Neolithic package seems to have been introduced by the first human settlers. In some respects, however, Mediterranean island ecosystems were quite different from adjacent continental ecosystems: the existence of endemic animal species is perhaps the most obvious example of this. On Sardinia/Corsica, Mallorca and Cyprus, the existence of such species gave rise to significant human adaptations: it is surely no coincidence that these were the only Mediterranean islands to be occupied by human communities which did not possess domesticated livestock. In the case of Mallorca, it seems clear that the earliest settlers came from a parent population on the Iberian mainland which did have domestic animals. The fact that they did not, in the initial stages, import livestock to the Balearic Islands is probably due to the abundance of the endemic caprine, *Myotragus* (Clutton-Brock 1984). Having evolved in an environment without predators, endemic species are likely to have had little fear of humans, and would thus be relatively easy to hunt or, in the case of *Myotragus*, to herd. From the ecological point of view, therefore, island environments may have their advantages as well as their disadvantages for human communities. Mediterranean prehistory offers numerous examples of human adaptations to insular environments which are distinctively different from mainland environments, but few, if any, examples of adaptations to severe resource limitation. This is perhaps unsurprising, given that none of the Mediterranean islands are particularly remote. It is also true to say that the vast majority of the Mediterranean islands were not occupied until after the transition to food production, and that those which were occupied by hunter-gatherer groups were the larger islands with the widest range of resources.

2   *We would also expect to see evidence for resource depletion and declining biodiversity on islands following colonisation.*

MacArthur's and Wilson's (1967) concept of 'species equilibrium' suggests that the colonisation of an island by a new species (including humans) is likely to be balanced by the extinction of species already there. We might expect this effect to be particularly marked in the case of human colonisations, since humans have historically been extremely successful competitors in the evolutionary arena. Endemic species, having evolved in the absence of predators, are likely to be especially vulnerable. There is empirical evidence from Pacific islands such as Tikopia (Kirch and Yen 1982) and Easter Island (Bahn and Flenley 1992) for significant resource depletion following human colonisation.

As we have seen, however, the impact of hunter–gatherer communities on the ecology of the Mediterranean islands seems in most cases to have been relatively limited. On Sardinia/Corsica, human communities seem to have coexisted with *Prolagus* for at least 7000 years, and perhaps for much longer. On the Balearic islands, human groups coexisted with *Myotragus* for around 3000 years. The exception to this pattern is Cyprus (Simmons 1991), where pygmy hippopotamus and elephant do seem to have been wiped out by human hunting. The essential difference between Sardinia/Corsica on the one hand and Cyprus on the other is probably to be found in the reproductive biology of the endemic species themselves: *Prolagus*, a relative of the rabbit, is likely to have avoided 'blitzkrieg' extinction (cf. Mosimann and Martin 1975) by rapid and prodigious breeding, whilst the pygmy elephants and hippos of Cyprus were unable to reproduce at a fast enough rate (Diamond 1989). The situation of *Myotragus* is rather different, in that its survival is probably due to deliberate herd management by human communities.

If the impact of hunter–gatherer communities on Mediterranean island ecosystems was limited, however, the same cannot be said of the impact of agricultural communities. Neolithic colonists brought domestic animals to the Mediterranean islands, which competed with endemic species for grazing/browsing, and probably also brought diseases to which these species had no immunity. The natural vegetation was cleared for cultivation and pasture, destroying the habitat of native species. *Prolagus* on Sardinia and Corsica and *Myotragus* on the Balearic islands became extinct, in a classic illustration of Diamond's (1989) 'sitzkrieg' model. In one sense, this pattern actually confirms MacArthur's and Wilson's (1967) model of 'species equilibrium': it could be argued that island ecosystems such as those on Sardinia/Corsica and the Balearic islands could accommodate one further species (humans), but could not accommodate the full package of new plant and animal species which arrived as part of the transition to food production. In another sense, however, it demonstrates the inadequacy of the island biogeography model in relation to human communities: the effect of human colonists on an island ecosystem depends not only upon their biology, but also on a range of cultural factors. The transition to food production on Corsica/Sardinia and on the Balearic Islands should probably not be seen as an environmental adaptation (there is no evidence to suggest that the previous food procurement strategies were unstable), but as a cultural strategy permitting the production of an increased food surplus, perhaps in order to fulfil social demands (cf. Bender 1978). For human communities, therefore, ecology is not independent of culture.

3   *Given that the 'carrying capacity' (cf. Williamson and Sabath 1984) of an island depends on the food procurement strategies of human communities, we might expect phases of island colonisation to coincide with significant thresholds, such as the transition to food production, or the 'secondary products revolution'.*

This is precisely the pattern which we identified in the previous chapter. The evidence from sites such as the Francthi Cave suggests that maritime technology and navigational knowledge were probably not significant barriers to the colonisation of the Mediterranean islands by human groups in the early Holocene, yet the majority of islands were not colonised until after the transition to food production. Carrying capacity can therefore be seen as the main barrier to human colonisation, i.e. in terms of MacArthur's and Wilson's (1967) model, the size of islands was probably more significant than their distance from the mainland. This would also explain why Sardinia/Corsica, Cyprus and the Balearic Islands were colonised at an early stage.

4   *We might expect Neolithic communities on the Mediterranean islands to adapt food-producing strategies to suit their own specific needs, giving rise to distinctive insular Neolithic packages.*

This is the phenomenon which Lewthwaite (1985a) has identified on Corsica, arguing for a 'forest Neolithic' in contrast to the normal Mediterranean cereal/ovicaprid regime. Certainly it seems that established human communities on islands such as Sardinia, Corsica and the Balearics were able to 'pick and choose' elements of the Neolithic package, leading to a 'filter effect' which, according to Lewthwaite's model, had a significant effect on the dispersal of the Neolithic package from Italy to southern France. This effect, however, only applies to the relatively few Mediterranean islands which were already colonised prior to the transition to food production. On islands such as Crete, which were colonised for the first time during the Neolithic, a ready-made package of food producing strategies were introduced (Broodbank and Strasser 1991).

5

ELABORATION AND CONTINUITY IN MEDITERRANEAN ISLAND PREHISTORY

It is widely accepted among biologists that island environments have particular ecological features which distinguish them from mainland environments. These insular characteristics are reflected in the range and nature of the animal and plant species found on islands, as noted by Darwin (1969 [1859]) and by MacArthur and Wilson (1967). Human colonists, like other animals, must either adapt to the unique insular ecosystem or become extinct. In the previous chapter, we explored a range of prehistoric human adaptations to island ecosystems in the Mediterranean. The question then arises, however, as to whether human island communities also have unique cultural characteristics. Evans (1973) argues that some island communities show tendencies towards the elaboration of particular elements of their culture, often associated with religious or ceremonial life: the Maltese temples and the 'Ahus' of Easter Island are obvious and spectacular examples of this phenomenon. Is it coincidence that these monuments occur on islands, or is there a specific sociogeographic explanation? There are many examples of cultural elaboration in the prehistory of the Mediterranean islands, not all of them necessarily associated with religion: the taulas and talayots of the Balearic Bronze Age, the nuraghi of Bronze Age Sardinia and the palaces of Minoan Crete are all specifically insular phenomena, and represent considerable investments of communal labour: why do such monuments not occur in mainland areas of Spain, Italy or Greece?

If island communities can, given the right combination of circumstances, be centres of innovative cultural elaboration, their remoteness and security from attack can also make them bastions of cultural conservatism. Particular cultural characteristics may survive in island communities long after they have disappeared from adjacent mainland areas. Thus pottery of distinctive Minoan type continued to be made on the island of Karpathos long after the collapse of Minoan Crete (Melas 1985), and elements of Mycenean civilisation survived into the Iron Age of Cyprus (Karageorghis 1982).

In this chapter, the issues of cultural elaboration and conservatism on islands will be explored through a series of Mediterranean case-studies, in an attempt to understand the variables which affect the development of these phenomena. The aim is not to suggest that insularity is either a necessary or a sufficient criterion for such developments (which would be absurd), but rather to explore the interaction of geographical and other factors in defining the cultural uniqueness of some island communities at particular points in time and space.

TAULAS AND TALAYOTS: MONUMENTAL ARCHITECTURE IN THE BALEARIC ISLANDS

The later prehistory of the Balearic Islands has been divided into the following phases (Plantalamor and Rita 1984):

- Pretalayotic (Chalcolithic/Early Bronze Age)   3400–1700 cal. BC
- Talayotic I (Middle/Late Bronze Age)   1700–1150 cal. BC
- Talayotic II (Late Bronze/Early Iron Age)   1150–800 cal. BC
- Talayotic III (Iron Age)   800–300 cal. BC
- Talayotic IV (Classical influences)   300–123 cal. BC

(N.B. Waldren (1982) considers Phases III and IV to be 'post-Talayotic').

The Pretalayotic period is further divided by Waldren (1982) into a 'Neolithic Early Ceramic phase' (3400–2450 cal. BC), an 'Early Beaker phase' (2450–2000 cal. BC) and a 'Late Beaker phase' (2000–1700 cal. BC). The material culture of the Pretalayotic period demonstrates clear links to the Iberian mainland: the material from Early Ceramic phase horizons at Son Matge has affinities in the mainland assemblages of Tabernus, Abrigio de Ambrosio and Cueva de l'Or, whilst the pottery of the Early and Late Beaker phases clearly belongs to the European Bell-Beaker tradition. Evidence from Son Matge also suggests that metalworking was introduced to the islands at around 2000 cal. BC (Waldren 1982). Most sites of the Pretalayotic period are open-air settlements, such as Ferrandell-Oleza: this site is a large walled village, incorporating a tower-like structure and other circular outbuildings. Some settlements, as at Son Mercer de Baix (Rita 1988; Plantalamor and Rita 1984) are characterised by boat-shaped houses or 'navetiformes', which seem to represent a uniquely Balearic architectural form. Burial sites of the Pretalayotic phase include megalithic tombs, which are found principally in the eastern part of Menorca, as at Roques Llises and Alcaidus, though there is one example, Son Baulo de Dalt, on Mallorca, and one, Ca na Costa, on Formentera. Most of these are rectangular cists, approached by a short passage with a porthole slab at the junction of passage and chamber (Rita 1988). Ca na Costa, however, is a circular passage grave with clear parallels on the Iberian mainland as, for example, at Los Millares (Topp *et al.* 1976). Rock-cut tombs are also found, mainly in the western part of Menorca, as at Son Vivo, Torre del Ram and Son Mercer de Dalt (Rosello-Bordoy 1963). Finds from both megalithic and rock-cut tombs include Bell-Beaker sherds, archers' wristguards, V-perforated buttons, copper awls and bronze daggers.

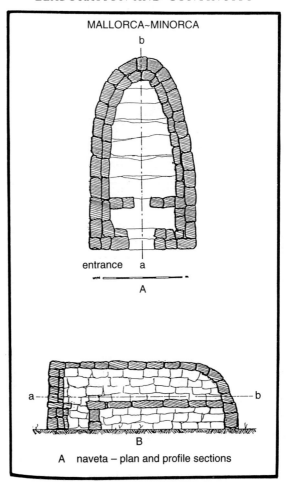

Figure 5.1 Structure of a typical naveta

Source: Waldren 1982

The Talayotic period, which begins at around 1700 cal. BC, is marked by the appearance of three monument types, all of which are unique to the Balearic Islands. The first of these monument types is the 'naveta'. A naveta is a drystone construction with the general appearance of an up-turned boat: in plan it may be horse-shoe shaped, oval or sub-circular (Figure 5.1). Pericot-Garcia (1972) and others have pointed out the resemblance in plan between the true navetas (which appear to have a funerary function) and the 'navetiforme' houses, found in villages such as Boquer, Sa Vall and Son Mercer de Baix. The excavations at Son Mercer de Baix (Rita 1988; Plantalamor and Rita 1984) have shown that the navetiforme villages have their origins in the Pretalayotic period, and it is likely that the true navetas represent a symbolic transformation of this uniquely

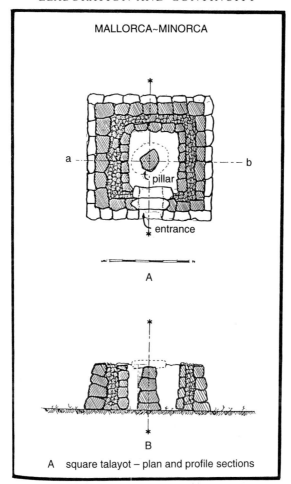

Figure 5.2 Plan and profile of a square talayot

Source: Waldren 1982

Balearic house form, and the creation of a 'house of the dead'. Waldren (1982) lists thirty-six known navetas, all on the island of Menorca (Mascaro-Pasarius (1958) reported sixty-four examples). This distribution, however, may simply reflect the conditions of preservation on the different islands: preservation on Mallorca is much poorer, owing to intensive cultivation in recent times. A typical naveta, such as Els Tudons (Serra-Belabre 1964), has two floors (Figure 5.1): the upper floor appears to have been used as a 'drying shed' for corpses (Waldren 1982), whilst the lower floor was used for the deposition of disarticulated bones, following the removal of flesh. The remains of more than 100 people were found at Els Tudons, together with V-perforated buttons and bronze bracelets, suggesting an Early Talayotic date (Serra-Belabre *op. cit.*). The

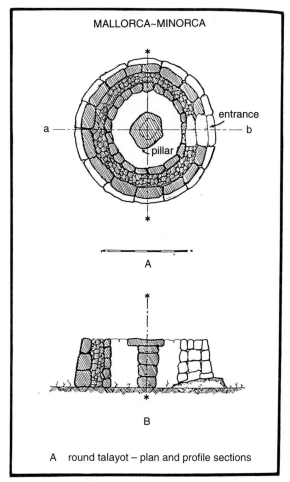

Figure 5.3 Plan and profile of a circular talayot

Source: Waldren 1982

tradition of building 'boat-shaped' tombs, however, seems to have persisted at least until the seventh century BC: the cemetery of Son Real includes a number of 'micro-navetas', containing up to six skeletons and dated to the seventh and eighth centuries BC (Pericot-Garcia 1972).

The second major monument type is the 'talayot' or 'watch-tower'. A talayot is a tower-like structure, which may be square (Figure 5.2), round (Figure 5.3), oval or stepped. Waldren (1982) argues for a chronological sequence, with the circular talayots being the earliest, followed by the square and stepped forms. The diameter of a talayot varies from 12 m–20 m, and the roof is in most cases supported by a massive central pillar (Figures 5.2 and 5.3). Talayots occur both on Mallorca and Menorca (Waldren (1982)

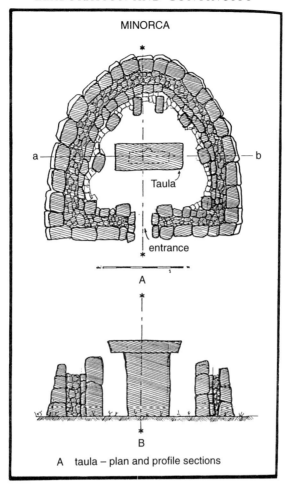

Figure 5.4 Plan and profile of a typical taula
Source: Waldren 1982

lists 205 on Mallorca and 134 on Menorca). Large-scale excavations have shown that talayots were, in most cases, integral features of settlements. At El Pedregar, for example, a walled settlement was found with over thirty houses: several talayots were built against the walls and within the settlement itself (Pericot-Garcia 1972). The settlements of Els Antigors and Ses Paisses were each built around a central talayot. The site of Ses Paisses (Lilliu 1960, 1962a) measures 94 m by 106 m, and is surrounded by a wall. The stratigraphy of this site suggests that occupation continued into the second and third centuries BC. Waldren (1982) draws attention to the large size of the Talayotic settlements: some, such as Catlar, Torre den Gaumes and Trepuco, being up to seven times larger than Troy.

The third monument type is the 'taula' or table. A taula consists of a horseshoe-shaped walled enclosure: the interior side of the enclosure wall is lined with orthostats, and at the centre of the enclosure is a massive pillar with a single capstone, forming a stone 'table' (Figure 5.4). With the exception of an atypical monument at Almallutx (Mallorca), taulas are known only from Menorca (Mascaro-Pasarius 1968). Although it has been suggested (Florit-Piedrabuena 1969) that the taulas are purely functional structures, and that the stone table was intended as a support for a roof, there is no convincing evidence to suggest that any of the taulas were roofed, and most authorities (cf. Pericot-Garcia 1972; Waldren 1982) consider that they functioned as sanctuaries. Several of the sites have produced large quantities of animal remains, suggestive of sacrifice and/or communal feasting. The recently excavated Taula Torralba d'en Sallort produced a series of radio-carbon dates calibrating between 1400 and 900 BC (Waldren 1982), though stratigraphic and artefactual evidence demonstrates continued activity on the site into the Roman period.

Our understanding of the monuments of the Balearic Bronze Age remains limited. As conspicuous features in the landscape, the majority have been plundered or dug into in antiquity. Few have been excavated with the benefit of modern techniques, and those that have have generally produced surprising results. The chronology outlined here (largely taken from the work of Rita (1988), Plantalamor and Rita (1984) and Waldren (1982) is imprecise, and by no means secure, as it is based on evidence from a very small number of excavated sites: it seems clear, however, that navetas, talayots and taulas overlap in date, and that the majority of them were built between *c*. 1400 and *c*. 800 cal. BC. As regards the function of these monuments, it seems that the true navetas (as opposed to the 'navetiform' structures) were funerary monuments, whilst the taulas were probably sanctuaries. The talayots must be seen as integral to large settlements, though the precise role of the towers themselves is unclear. Waldren (1982) suggests that they may have served 'some social or religious function', whilst Gasull *et al.* (1984) suggest (on the basis of the archaeological evidence from Son Fornes) that they served as a focus for ceremonial feasting for the inhabitants of the villages which surrounded them.

Navetas, talayots and taulas are expressions of a uniquely Balearic Bronze Age culture: none of these monument types has any convincing mainland parallels. There are superficial similarities between the Balearic talayots, the Corsican torri and the Sardinian nuraghi, but these similarities are not sufficiently close to suggest any direct links. The material culture of the Balearic Bronze Age is in any case quite different from that of Sardinia or Corsica, and has clear parallels on the Iberian mainland. The cultural uniqueness of the Balearic Bronze Age and Early Iron Age stands in contrast to both the Pretalayotic period and the Punic/Classical period, in which insular characteristics are expressed to a much lesser extent (though there are some unique features, such as 'navetiforme' houses in the Pretalayotic period, or quick-lime burials in the Iron Age). The Talayotic period, however, was by no means a period of isolation for the Balearic Islands. Insular uniqueness was expressed mainly through architecture (and probably, by implication, through the religious beliefs and ritual practices associated with the navetas and taulas), and is scarcely identifiable at all in the material culture, which has very clear mainland parallels. Waldren (1982) draws attention to the large quantity of bronze artefacts found in Talayotic contexts: since the necessary raw materials are not present in the Balearic Islands, this abundance must reflect exchange or trade with mainland communities. Waldren (*op. cit.*) also points out, however, that the decline of Talayotic building coincides chronologically with increasing trade and contact abroad, and with the availability of more varied and richer trade items, largely as a result of the founding of a Carthaginian colony on Ibiza in 654 BC. The pattern may become clearer when we look at Balearic prehistory as a whole, rather than focusing exclusively on the Talayotic period. The principal features of the culture sequence can be summarised as follows.

Early settlement period (*c.* 5500–3400 cal. BC)

This period is marked by the first human settlement of the Balearic Islands. Human communities on the islands do seem to have been largely isolated from those on the Iberian mainland, and had an entirely unique economy (based on hunting and herding of *Myotragus balearicus*) and material culture (aceramic).

Pretalayotic period (*c.* 3400–1700 cal. BC)

During the Pretalayotic period the islands became integrated into the eastern Iberian cultural area, adopting mainland economic practices (cereal cultivation and introduced domestic livestock), burial traditions and material culture, alongside some unique cultural expressions (e.g. 'navetiforme' houses). Island/mainland exchange is evident, particularly during the Early and Late Beaker phases, when exchange seems to have involved the circulation of prestige items, usually deposited in burials (beakers, archers' wrist-guards, copper daggers and awls, V-perforated buttons).

Talayotic I-II period (*c.* 1700–800 cal. BC)

This is the period of greatest cultural uniqueness, with the construction of navetas, talayots and taulas. Links with the Iberian mainland, however, are evident from material culture, and also from the large quantity of imported metalwork. Although exchange did not decline during this period, it is possible that its social significance was reduced, and that competitive social relations were no longer articulated through the control of access to prestige goods from the mainland. The number and size of talayotic settlements suggests a significant increase in the population of the islands during this period (Waldren 1982).

Talayotic III–IV (or post-Talayotic) period (*c.* 800–123 cal. BC)

This period sees the decline of Talayotic building (though some monuments remain in use, as at Taula Torralba d'en Salort), coincident with the rise of Punic, and later Roman, influence in the area. The establishment of the Punic colony on Ibiza may have given a new impetus to trade in the mediation of social status.

Thus a cycle can be identified, with the Pretalayotic and Talayotic III–IV periods reflecting 'exchange-oriented' societies, and the Talayotic I–II period reflecting a 'monument-oriented' society (cf. Stoddart *et al.* 1993). This does not imply a decline of trade in the Talayotic I–II period (which would not be compatible with the archaeological evidence), but rather a decline in the significance of trade in the establishment and maintenance of power relations. This is a theme to which we will return at a later stage.

NURAGHI AND TORRI: STONE TOWERS OF SARDINIA AND CORSICA

From a biogeographic perspective, Sardinia and Corsica are very different from the Balearic Islands. For one thing they are much larger: Sardinia has a surface area of 24,089 km^2, compared with Mallorca's 3740 km^2. Second, Corsica and Sardinia are less remote than the Balearics: the longest single sea journey required to reach Sardinia from Italy is 58 km, compared with 92 km to reach Mallorca from Spain. These factors combine to give Sardinia a T/DR Ranking (see Chapter 3) of 1.5, compared with 0.2 for Mallorca. We might reasonably expect these differences to be reflected in the culture sequences of the two island groups: specifically, one might expect Corsica and Sardinia to be less 'insular', and more integrated within the processes of cultural change observed on the Italian mainland.

Throughout most of the Neolithic period, Sardinia and Corsica were clearly tied in to broader regional patterns of exchange and interaction. Monte Arci, in Sardinia, was one of the most important sources of obsidian in the western Mediterranean, and Sardinian obsidian has been found extensively in Italy and in southern France (see Chapter 6). Neolithic cultural practices on the islands in many respects mirrored those on the Italian mainland: the burial evidence from the Early Neolithic Filiestru phase and the Middle Neolithic Bono Ighinu phase in Sardinia, for example, is dominated by cave burials and rock-cut tombs, as it is in peninsular Italy during the same period (Balmuth 1992; Whitehouse 1972). Burial sites are rarer in Corsica, but

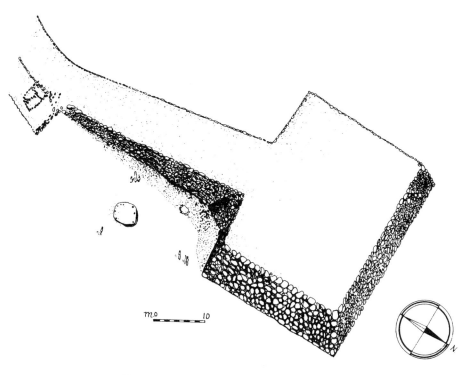

Figure 5.5 Reconstruction of Monte d'Accodi

Source: Contu 1984

rock-cut tombs are recorded in the Sartenais area in the southeast of the island, and one cave burial site is known at San Simeone (Lewthwaite 1984b). The specific cultural identity of the island communities, however, was expressed through material culture: the pottery styles of the Bono Ighinu phase and the Late Neolithic Ozieri phase in Corsica are distinctively insular in character (Balmuth *op. cit.*).

The beginning of the Ozieri phase marks a significant in the relationship of Sardinia to the rest of the west Mediterranean world. This phase is dated by a series of radiocarbon determinations from the sites of Sa'Ucca de Su Tintirriolu and Grotta Filiestru (Trump 1984):

Sa'Ucca de Su Tintirriolu
    4850 ± 50 bp = 3630–3540 cal. BC (R-789)
    4930 ± 50 bp = 3800–3690 cal. BC (R-883)
    5090 ± 50 bp = 3980–3800 cal. BC (R-884)
Grotta Filiestru
    4950 ± 50 bp = 3780–3690 cal. BC (Q-3028)
    5250 ± 60 bp = 4220–3690 cal. BC (Q-3027)

The Ozieri phase is marked by the appearance of new types of ritual site. The cave of Sa'Ucca de Su Tintirriolu has itself been interpreted (Trump *op. cit.*) as a ritual site. Access is particularly difficult, and the assemblage includes figurines and pottery associated with dancing figures, as well as burials. The ritual use of caves is a phenomenon well attested in Neolithic Italy (Whitehouse 1992), and is by no means unique to Sardinia. Other monuments of this period, however, are more unusual: the megalithic monument of Pranu Mutteddu (Balmuth 1992) consists of three concentric stone circles surrounding a series of tombs. It would be difficult to find a parallel for this monument on the Italian mainland or elsewhere (the superficial resemblance to the Brochtorff Circle on Gozo is presumably coincidental). Even more remarkable is the site of Monte d'Accoddi (Contu 1984), a massive stone platform, surmounted by a rectangular shrine and approached by a ramp, 75 m long by 37.5 m wide (Figure 5.5). The emergence of these monuments suggests the embryonic development of an insular religious tradition.

The succeeding Chalcolithic and Early Bronze Age periods are characterised by the appearance of a number of different 'cultures', whose relationship to one another is unclear (Balmuth 1992). In fact, many of these 'cultures' (Filigosa, Abealzu, Monte Claro, Bonnanaro) are simply pottery styles. The clearest indications of the chronological relationships between them are from the excavated cave settlement of Grotta Filiestru (Trump 1984). The stratigraphy of this site was as follows:

Context 6	Sa Turricula or Bonnanaro B pottery.
Context 5	Monte Claro and Bonnanaro pottery, also Beaker sherds and Beaker derived wares.
Context 4	Ozieri pottery
Context 3	Bonu Ighinu pottery
Context 2	Filiestru pottery
Context 1	Cardial pottery

Two radiocarbon dates from the Ozieri horizon have been quoted above. Two dates were obtained from context 5:

4430 ± 50 bp = 3290–2990 cal. BC (Q-3029)
3805 ± 40 bp = 2320–2140 cal. BC (Q-3030)

A single date was also obtained from context 6: 3440 ± 40 bp = 1870–1690 cal. BC (Q-3031).

The Early Bronze Age is apparently marked by the appearance of the megalithic tombs known as 'tombi di giganti': Bonnenaro pottery was found in the tomb of Goronna (Balmuth 1984). Tombi di giganti are rectangular megalithic chambers, with an entrance at one end, formed by a carved stone slab. The entrance is generally set at the centre of a crescentic forecourt, defined by upright stones (Figure 5.6). Some of the monuments contain large numbers of burials: sixty from Las Plassas, fifty from Preganti, thirty from

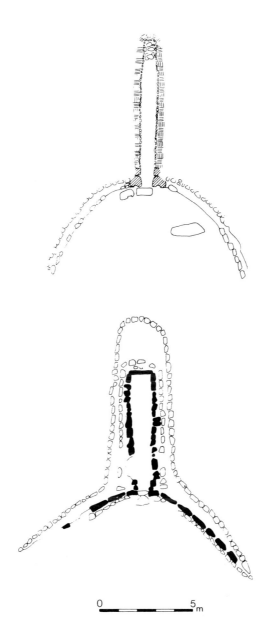

Figure 5.6 Plans of the tombi di giganti of Oridda and Li Mizzani
Source: Whitehouse 1981

Scusorgiu (Balmuth 1984). The ceramic assemblage from Goronna suggests an origin for these monuments in the late third millennium cal. BC, though some of the monuments contained bronze daggers, spears and swords, suggesting continued use into the Middle and Late Bronze Age. Other gallery graves, lacking the crescentic forecourt characteristic of the tombi di giganti, probably date to the same period: Bonnanaro pottery was found, for example, in the gallery grave of Ena'e Muros, along with two flat daggers. A series of simple dolmens, concentrated in the northwest of Sardinia, may date to the same period, but there is little clear dating evidence. The fifty or so recorded 'menhirs' in Sardinia are similarly difficult to date (though Castaldi (1984) points out that 'numerous fragments' of menhirs were re-used in the construction of a tombo di giganti at Aiodda-Nurallao). Many of the Sardinian menhirs have an anthropomorphic form, and stylised carved breasts (Atzeni 1980). At the site of Biriai, a setting of thirteen menhirs were found associated with three concentric walls running around a natural outcrop, at the centre of a prehistoric village: most of the pottery found on the site is of Monte Claro type (Castaldi 1984) suggesting a date in the third millennium cal. BC. Many of the Corsican standing stones have more explicitly carved anthropomorphic features, with helmets and swords depicted (Lewthwaite 1984b): again, dating is difficult, but the re-use of such statues in the 'torre' of Filitosa suggests a relatively early date within the Corsican Bronze Age.

The best-known monuments of the Sardinian Bronze Age, however, are the 'nuraghi', massive stone towers, superficially similar to the talayots of the Balearic Islands. The 'torri' of Corsica, though less well understood than the nuraghi, are clearly related to them. The classic 'tholos nuraghe' is a truncated conical tower of large masonry, measuring 11 to 16 m in diameter and up to 18 m in height (Lilliu 1962b). Some have two, or even three storeys, as at Santa Barbara-Macoma, Orolo-Bortigali and Madrone-Silanus (Gallin 1989). These nuraghi have a single entrance, leading to a central chamber, with a corbelled vault (Figure 5.7). The chamber may have niches set into the wall, as at Nuraghe Toscono-Borore (Webster 1991). Where upper levels exist, these are approached via a spiral staircase, leading up from the entrance passage. A second form of nuraghe, with a flat roof and bisected by a corridor leading to side chambers, was until recently considered to be a later development. The excavation of one of these 'corridor nuraghi' at Brunku Madugai, however, produced remarkably early evidence, including a radiocarbon date calibrating between 2070 and 1570 BC (Balmuth 1992). These structures are referred to as 'protonuraghi' by Demurtas and Demurtas (1984), who date them between 1800 and 1500 bc (i.e. c. 2170–1760 cal. BC). Trump (1984), however, has reported pottery of Sa Turricula type in a horizon associated with the construction of a tholos-type monument, Nuraghe Noeddos. He records that this pottery is identical to that from Grotta Filiestru, in a horizon which produced a radiocarbon date calibrating

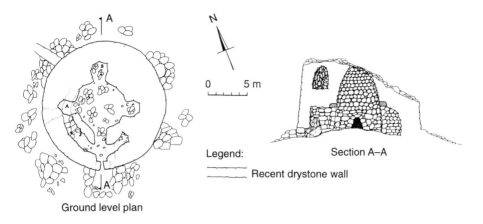

Figure 5.7 Plan of Nuraghe Toscono-Borore
Source: Webster 1991

between 1870 and 1690 BC (see above). It is possible, therefore, that both 'corridor' and 'tholos' nuraghi have their origins in the early part of the second millennium cal. BC. The Corsican torri may have an even earlier origin: Lewthwaite (1984b) quotes the following radiocarbon dates from Torre Tappa:

4168 ± 110 bp = 2910–2580 cal. BC (Gif-94B)
3865 ± 125 bp = 2560–2140 cal. BC (GSY-94B)

The construction and use of nuraghi continued up to Roman times. Trump (1984) reports Roman period additions to Nuraghe Noeddos. During the 1500 years or so of Nuraghic civilisation, the monuments themselves underwent significant evolution. Four main phases of construction have been identified at Nuraghe Su Nuraxi, one of the largest surviving nuraghi. Tore (1984) distinguishes between 'simple nuraghi' and 'complex nuraghi'. Simple nuraghi are monuments such as Nuraghe Noeddos (Trump 1984), Nuraghe Toscono-Borore (Figure 5.7) and Nuraghe Monte Idda di Posada (Fadda 1984). These have a single tower, and seem to be the earliest nuraghi: the radiocarbon dates from Nuraghe Noeddos are quoted above, and the excavations at Nuraghe Monte Idda di Posada suggest construction between the sixteenth and the thirteenth centuries cal. BC. Complex nuraghi (Tore 1984), such as S'Urahi, Palmavera and Su Nuraxi (Figure 5.8) are far fewer in number. These are characterised by the presence of several large towers (S'Urahi has seven), linked together by walls. All the indications suggest that these are later than the simple nuraghi. Tore (1984) has demonstrated stratigraphically that the central tower at S'Urahi predates the rest of the complex: this

101

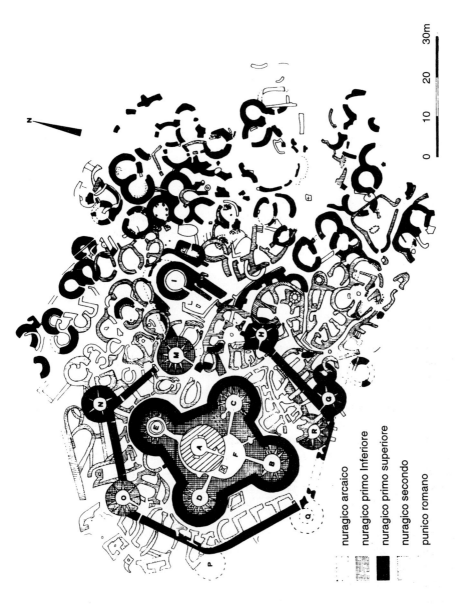

nuragico arcaico

nuragico primo Inferiore

nuragico primo superiore

nuragico secondo

punico romano

Figure 5.8 Plan of nuraghic complex of Su Nuraxi
Source: Lilliu 1962b

monument, therefore, began as a simple nuraghe, and was elaborated at a later stage, probably between 900 and 500 cal. BC. Many nuraghi, particularly the larger complex examples, are surrounded by settlements, as at Serucci (Balmuth 1992) and Su Nuraxi (Balmuth 1984). The nuraghic complex of Serucci produced a series of radiocarbon dates (Balmuth 1992):

2710 ± 45 bp = 900–830 cal. BC (PITT-0516)
2930 ± 50 bp = 1250–1040 cal. BC (PITT-0517)
2795 ± 30 bp = 990–920 cal. BC (PITT-0518)

The Corsican torri seem never to have reached a level of elaboration comparable to that of the complex nuraghi (Lewthwaite 1984b), and it is unclear to what extent their construction and use continued into the first millennium cal. BC.

Other features of the later nuraghic period (first millennium cal. BC) include a unique series of copper figurines, depicting warriors (sometimes with horned helmets), animals and, in some cases, boats. These are quite unlike contemporary Greek or Villanovan bronzes from the Italian mainland. These are sometimes found, associated with miniature bronze swords and other weapons, as votive deposits in sacred wells, as at Santa Vittoria di Serri. These 'well-sanctuary' sites (Lilliu 1975) consist of an enclosure built around a well, with a covered staircase descending to the water level: the entrance to the staircase is generally set within a courtyard, defined by a low stone wall.

Like the talayotic culture of the Balearic Islands, the nuraghic culture of Sardinia and associated torrean culture of Corsica represent the expression of clearly defineable insular traditions. As in the Balearic Islands, the emergence of this monumental tradition can perhaps be linked to a decline in the social importance of island/mainland exchange (in this case the exchange of Sardinian obsidian), whilst its decline coincides with the increasing influence of Classical civilisation. Lewthwaite (1985b) suggests that: 'the autonomous episode of the Nuraghic Middle Bronze Age appears . . . as an involuted phase . . . sandwiched between the periods of intense inter-regional interaction'. As regards the nuraghi themselves, they have been variously interpreted as fortresses, watch towers, meeting places, religious buildings and prestige symbols (Balmuth 1992). Perhaps the most influential interpretation has been that of Lilliu (1959b, 1962b), who views them as 'protocastles', the defended residences of a quasi-feudal elite. Webster (1991) points out that excavations have provided little evidence for the aristocratic households, professional soldiers or occupational specialists predicted by the 'feudal' model: he prefers a model of small-scale polities, analogous to African petty chiefdoms (cf. Taylor 1975), with each nuraghe serving as the fortified residence of a chiefly household. Webster (*op. cit.*) estimates the total labour expenditure required to build a simple tholos nuraghe at 3600 labour days, compared with 18,000 for a large complex nuraghe. This is a relatively modest labour investment (significantly less, for example, than would be required to build many of the Neolithic monuments in temperate Europe) and need not imply the existence of a state society. The appearance of the large complex nuraghi in the first millennium cal. BC, however, does represent the emergence of a distinct settlement hierarchy, suggesting a degree of sociopolitical centralisation which cannot be overlooked. Most scholars (Balmuth 1984; Lilliu 1962b; Webster 1991) assume that the function of the nuraghi was essentially domestic. Whilst it is certainly true that many nuraghi are surrounded by villages of round houses there is, in most cases, very little archaeological evidence for the function of the nuraghi themselves. The Corsican torri, which are clearly related to the nuraghi, are often considered to have had a religious significance: the access passage is generally narrow and restricted, and the incorporation of reused statue menhirs in the construction of the torre of Filitosa is clearly suggestive of a religious dimension. Whatever the function of the nuraghi and torri, whether as defended residences, village watchtowers or religious sanctuaries, their size and construction suggest that they were buildings of considerable symbolic importance, reflecting the power of an emergent elite grouping. That the power of this elite was based, at least in part, on the control of religion, is suggested by the effort expended in the construction of tombi di giganti and well sanctuaries, and by the wealth deposited in the latter in the form of votive offerings. This is yet another point of comparison with the Balearic case, where comparable effort was invested in the construction of taulas and navetas. At all levels, the comparisons between the Talayotic culture of the Balearics and the Nuraghic and Torrean cultures are striking, yet there is no good evidence for a direct connection between them. Not only is the material culture of the Balearics different from that of Corsica and Sardinia, the monuments themselves are actually quite different. A Balearic naveta and a Sardinian tombo di giganti are both burial monuments, but they are quite different in design and conception. The differences between a Balearic taula and a Sardinian well sanctuary are even more marked, though both can probably be seen as religious or cult sites. Perhaps, then, the similarities between these two case-studies should be understood in sociogeographical rather than in culture-historical terms: could it be the 'island effect', rather than any direct cultural link, that explains these similarities?

THE STONE TEMPLES OF MALTA AND GOZO

Malta and Gozo are among the most remote islands in the Mediterranean: they are not directly visible from any mainland (though it is said that, on occasions, the fiery summit of Mount Etna can just about be seen on the horizon from Malta) and require a sea voyage of 80 km to reach them from Sicily. Their small size and relative remoteness give the islands a T/DR ranking (see Chapter 3) of 0.1, one of the lowest rankings of any Mediterranean island.

The Maltese Islands have attracted a great deal of attention in the archaeological literature (Zammit 1930; Evans 1959, 1971a) on account of the remarkable stone temples. Following on from Zammit's (1930) early work, Trump (1966) carried out important excavations at the temple site of Skorba. More recently, the culture sequence of Maltese prehistory has been revised (Evans 1984) and recent excavations at the Brochtorff Circle (Bonnanno *et al.* 1990) have given rise to new ideas about the social and cultural significance of monumental ritual in the Maltese Neolithic (Stoddart *et al.* 1993). As a result of this research it is now possible to outline, with some confidence, a detailed chronology of Maltese earlier prehistory. The principal features of the culture sequence can be summarised as follows (Stoddart *et al.* 1993).

Ghar Dalam phase (*c.* 5000–4500 cal. BC)

This period marks the earliest evidence for human settlement on the Maltese Islands. The pottery has clear affinities with the Stentinello style of Sicily, suggesting Sicily and southern Italy as the most likely staging point for the colonisation of the islands.

Skorba phase (*c.* 4500–4100 cal. BC)

The Skorba phase is marked by the earliest evidence for ritual activity on the islands: a small shrine at Skorba (see below), which should almost certainly be seen as a precursor of the later stone temples. The material culture of this period shows continued close links with Sicily and southern Italy.

Zebbug phase (*c.* 4100–3800 cal. BC)

Again, the evidence suggests continued mainland links. The dark incised pottery characteristic of this period has close similarities to the San Cono-Piano Notaro style of Sicily. Exchange items include obsidian from both Lipari and Pantelleria, ochre from Sicily and stone axes from Calabria. There is further evidence for cultural divergence in ritual practice, in the form of anthropomorphic pendants from Brochtorff and small statue menhirs from Brochtorff and Zebbug.

Ggantija phase (*c.* 3600–3000 cal. BC)

The Ggantija phase is marked by the appearance of the stone temples. Alongside the temples themselves are funerary sites (Hal Saflieni and the Brochtorff Circle) which can, in some senses, be seen as underground transformations of the temples themselves. Although there is some evidence (from Skorba) to suggest that obsidian was still in circulation, the evidence does suggest a marked reduction in the importance of exchange between Malta, Sicily and the Italian mainland.

Tarxien phase (*c.* 3000–2500 cal. BC)

The Tarxien phase is marked by the further, spectacular elaboration of the temples and associated funerary sites. Stoddart *et al.* (1993) have pointed out that this is also one of the most isolated phases in Malta's prehistory, with little evidence for external trade or contacts.

Tarxien Cemetery phase (*c.* 2500–1500 cal. BC)

This period is marked by the abandonment of the stone temples and the appearance of a new burial tradition (cremation cemeteries). It is also marked by the appearance of new material culture elements, including bronze tools and weapons, and a new pottery style with affinities in the Capo Graziano style of Lipari, and the Moarda style of northeastern Sicily. Whilst these changes have traditionally been interpreted as evidence for an invasion (Evans 1959; Trump 1976), the most recent work (cf. Bonnanno *et al.* 1990) has tended to favour internal processual explanations.

Already some familiar patterns are emerging: in particular, the inverse relationship which seems to exist between the importance of island/mainland exchange and the development of insular monumental traditions. In the Maltese case this is even more evident than in the Balearic Islands or in Corsica and Sardinia, since exchange dries up almost completely in the Ggantija and Tarxien phases. Stoddart *et al.* (1993) suggest that: 'the rivalry between families in pursuing exchange outside the Maltese Archipelago in the Zebbug Phase was transferred in the Ggantija Phase to rivalry between families in the construction of temples'. It may well be that this reflects a more general phenomenon in island sociogeography, an idea to which we will return in the final chapter of this book.

One thing that has become increasingly clear from the research carried out in Malta and Gozo over the past twenty years, is that the processes of cultural change which can be identified from the archaeological record happened relatively gradually. This discredits earlier notions that each cultural change corresponds to the arrival of a new group of people.

The 'shrine' discovered in the Skorba-phase horizon at the site of Skorba itself (Trump 1966) must surely be seen as a precursor of the temples. This building formed part of a settlement, stratified beneath the Ggantija Phase temple of Skorba. The 'shrine' (Figure 5.9) is the largest building in the village: the main room is oval in shape, and has a D-shaped room lying to the south of it. To the east and west are stone-paved courtyards. In the north room were found five female figurines, one in stone, the others in clay, together with six mutilated goat skulls and a number of artificially polished cattle tarsals. The earliest true temples, belonging to the Ggantija phase, are relatively simple constructions. That at Skorba is fairly typical (Figure 5.10), with a D-shaped plan and concave facade, enclosing a three-lobed structure approached by a passage. During the Tarxien phase the Skorba monument, like most of the Maltese temples, was extensively modified (Figure 5.11). The central apse was closed off by an orthostatic wall, with a massive trilithon doorway. Two altars were added inside the apse, and two more outside. The side apses were also marked off by the addition of steps (Trump 1966). A second temple was added to the east of the original monument, consisting of four apses in two opposed pairs. The modifications which occured at Skorba, as at other sites in the Tarxien phase, can be understood as an attempt to restrict access to the ceremonies conducted within the monuments: areas which had been open and visible (e.g. the central apse at Skorba) became closed in and hidden. Stoddart *et al.* (1993) point out that the main concentrations of cult objects and images are often found in the innermost recesses of the temples. At Tarxien, for example, apses 10 and 15, together with the so-called 'oracle room' have major concentrations of stone bowls, carved images and figurines of obese women: these apses also have door jambs, complete with holes for ropes or bars, allowing them to be closed to the outside. Apse 6 at Tarxien contained a midden of animal bones, suggesting sacrifice (Stoddart *et al.* 1993), whilst the central niche of the East temple at Skorba contained the skeleton of a large bird and an inverted stone bowl (Trump 1966). The changes which occured in Maltese temple architecture during the Tarxien phase suggest that access to the sacred domain was increasingly controlled by an elite. Meillassoux (1964, 1967) has argued that social power in tribal societies is often based on control over sacred knowledge and ritual practice (particularly initiation rites), and the Maltese evidence fits well within this model. The precise nature of the controlling elite in the Maltese case is open to debate. Renfrew and Level (1979) see the temples as territorial centres, each serving a discrete social group: as population rose, so these groups came increasingly into competition with one another. Inter-group competition was articulated through the building of larger and larger temples, associated with the emergence of a centralised hierarchy on 'chiefdom' lines (cf. Sahlins 1963). Bonnanno *et al.* (1990) point out that the temples do not occur as individual monuments (as territorial centres might be expected to) but in clusters or pairs, suggesting an element of intra-group as well as inter-group competition. Following Boissevain (1974), they argue that a centralised hierarchy is unlikely to develop in a small island community, characterised by dense social networks (Boissevain argues that in communities of less than 3000 to 4000 people, social relations are generally articulated through kinship and personal contacts, rather than through impersonal and rigid hierarchies: Renfrew (1974) estimates the prehistoric population of Malta to be 1000 to 2000 for each temple group). The model developed by

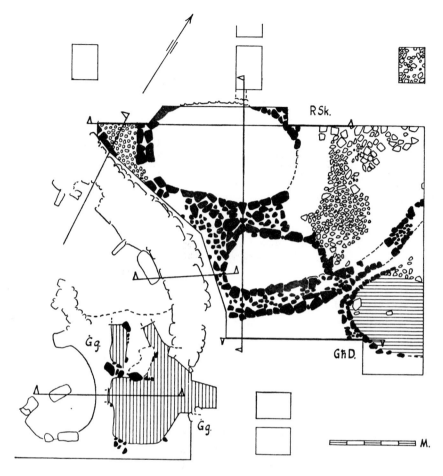

Figure 5.9 Plan of the early shrine at Skorba

Source: Trump 1966

Bonnanno *et al.* 1990 involves prominent individuals collecting personal factions 'recruited on the basis of broadly similar but rival ideologies': this comes closer to Sahlins' (1963) concept of a 'Big Man' system.

Whatever the nature and composition of the elites responsible for the Maltese temples, it seems likely that they controlled ceremonies and rituals connected with death, as well as those associated with other stages of the life cycle. The funerary complexes of Hal Saflieni and Brochtorff are geographically associated with temples (Brochtorff with Ggantija and Hal Saflieni with Tarxien). Although the temples are above ground and the funerary complexes below ground, there are a number of architectural and iconographic features

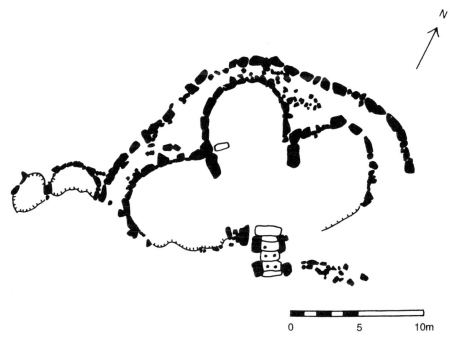

N

0   5   10m

Figure 5.10   Plan of Ggantija phase temple at Skorba

Source: Trump 1966

which link them (Stoddart *et al.* 1993). Both temples and funerary complexes have an essentially modular structure. In the case of the temples, the modules are apsidal structures containing particular ritual features, such as altars and stone statues. In the case of the funerary complexes, the modules are niches and rock-cut cavities, leading off from a central area. At Brochtorff (Stoddart *et al. op. cit.*), one of the excavated cavities was filled with both articulated and disarticulated human remains, associated with small terracotta figurines, presided over by a stone-skirted statuette. The figurines and statuettes are of obese women, closely comparable to those found in the temple iconography. The picture that emerges of the social structure of the Maltese Islands during the Ggantija and Tarxien phases, therefore, is one of competing small-scale elites, whose power was mediated through control of access to the sacred sphere represented by the temples and funerary complexes. This control, which seems to have become increasingly institutionalised, judging from the architectural changes which occured during the Tarxien phase, probably included the conduct of important rituals associated with life and death.

The abandonment of temple architecture in the Tarxien Cemetery phase must represent a significant change in this social order. The construction of

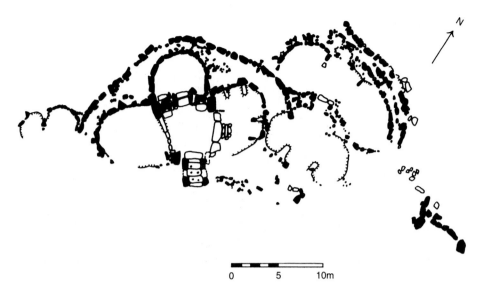

Figure 5.11 Plan of Tarxien phase temple at Skorba
Source: Trump 1966

monumental architecture ceased completely, although the use of existing monumental sites continued in a transformed way: the temple at Tarxien, for example, was transformed into a cremation cemetery (Zammit 1930). The iconography associated with the temples (most notably the skirted obese women) disappeared entirely, as did the funerary practices associated with sites such as Hal Saflieni and Brochtorff. Zammit (1930) noted the existence of a sterile deposit between the Tarxien phase and the Tarxien Cemetery phase deposits at the site of Tarxien, and argued for a long period of abandonment before the temple was finally reused as a cremation cemetery. Evans (1953), however, rejects this argument on typological grounds, since he considers the material culture of both the Tarxien phase and the Tarxien Cemetery phase to have affinities in the (relatively short-lived) Castelluccio culture of southeastern Sicily. Both Zammit (1930) and Trump (1976) stress the apparant lack of continuity between the Tarxien phase and the Tarxien Cemetery phase, and argue that the latter must reflect the arrival of a new group of settlers from Sicily and/or the Italian mainland. Bonnanno *et al.* (1990), however, argue for a social transformation, associated with a new form of ritual expression. It is also worth noting that the material culture of the Tarxien Cemetery phase, including the appearance of imported metal objects, suggests a renewed emphasis on exchange between the Maltese Islands, Sicily and the Italian mainland. It may be, therefore, that we can identify a shift from a 'monument-oriented' society back to an 'exchange-oriented' society, similar to that which had existed in the Skorba and Zebbug phases (Stoddart *et al.* 1993).

ELABORATION AND CONTINUITY IN THE AEGEAN BRONZE AGE

The Aegean Bronze Age lasted from around 2700 cal. BC to around 1200 cal. BC, and is conventionally subdivided into three broad phases. The earliest phase (Early Helladic I–III in mainland Greece, Early Minoan I–III in Crete and Early Cycladic I–IIIA in the Cyclades) lasts from *c.* 2700 to *c.* 2200 cal. BC, the middle phase (Middle Helladic in mainland Greece, Middle Minoan I–IIIA in Crete and Early Cycladic IIIB/Middle Cycladic in the Cyclades) from *c.* 2200 to *c.* 1700 cal. BC, and the late phase (Late Helladic, Early Minoan IIIB/Late Minoan and Late Cycladic) from *c.* 1700 to *c.* 1200 cal. BC (Barber 1987, Dickinson 1994). This period of around 1500 years saw the rise and fall of Europe's first literate 'civilisations' in Minoan Crete and Mycenean Greece. Island communities, most importantly Crete, played a key role in these processes, and the palaces of the Middle and Late Minoan periods can be considered as further examples of insular elaboration (the 'palace-based civilisation' of Crete evolved several centuries before that of the Greek mainland). Because of the complexity of the evidence from this period, it is convenient to look separately at the Early, Middle and Late Bronze Ages.

The Early Bronze Age (*c.* 2700–2200 cal. BC)

Crete and the Cyclades appear to have been settled at an earlier stage than the rest of the Aegean Islands: an important series of radiocarbon dates from Knossos (see Chapter 3) suggest that Crete was colonised in the early seventh millennium cal. BC, whilst the evidence from the Cyclades suggests colonisation in the fifth and sixth millennia cal. BC. Communities on these islands, therefore, were already well established by the beginning of the Bronze Age, particularly on Crete where, in contrast to many of the smaller Cycladic Islands, settlement appears to have been permanent from the time of initial colonisation.

There is little in the Early Bronze Age evidence to suggest a special role for Crete. Significant population centres at Knossos, Phaistos, Mallia and Mochlos in Early Minoan I (*c.* 2700–2600 cal. BC) have been inferred from the spread of material, and from the size of the cemeteries (Whitelaw 1983), and of these, Knossos is by far the largest, covering around 5 Ha. Evidence from the excavated sites of Mochlos and Ellenais Artaniou (Renfrew 1972) suggest that these settlements consisted of scattered rectangular houses. The number of settlements seems to have increased significantly in Early Minoan II (*c.* 2600–2300 cal. BC), and this was matched by an increase in architectural complexity: the

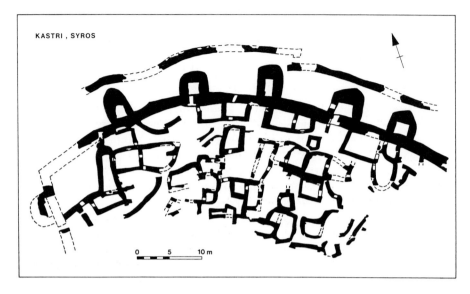

Figure 5.12   Fortified settlement of Kastri, Syros

Source: Bossert 1967

'house on the hill' at Vasiliki (Zois 1976), for example, had a paved courtyard, as did the EMII structure found beneath the Western Court at Knossos (Dickinson 1994). A similar pattern of development can be identified in the Cycladic Islands (Barber 1987). The settlement of Grotta (Naxos), dating to Early Cycladic I, has scattered rectangular houses similar to those recorded at the contemporary Cretan sites of Mochlos and Ellenais Artaniou, whilst the Early Cycladic II site of Aghia Irini (Keos) has more substantial architecture. In Early Cycladic III (2300–1800 cal. BC) fortified sites appeared, as at Kastri (Syros: Bossert 1967 and see Figure 5.12), Panormas (Naxos) and Mount Kythnos (Delos). This, however, is part of a generalised process of settlement expansion, nucleation and increasing complexity which can be identified across the Aegean area, and which is by no means restricted to the islands. The 'corridor villas' of Early Helladic II (*c.* 2600–2300 cal. BC) on the Greek mainland, as at the 'House of the Tiles', Lerna, and Weisses Haus, Aegina, are larger and more elaborate than any contemporary buildings in Crete or the Cyclades.

The Early Bronze Age is also marked, across the Aegean area, by the appearance of new burial practices. The burial record in the Cyclades is dominated by cist graves, which appear in Early Cycladic I (Barber 1987), whilst in Crete, circular 'tholos' tombs are present, as at Lebena (Branigan 1970), at a similarly early date. An important feature of the cist graves of Early Cycladic I-II is the presence in some of marble figurines, clearly made in the

112

Cycladic Islands themselves. It seems clear that these figurines reflect the elaboration of an insular tradition, the roots of which can be traced back into the Neolithic: a schematic figurine from Saliagos and a series of terracotta heads from Kephala can be securely dated to the Later Neolithic (Renfrew 1968). The raw materials for these figurines are present on the Cycladic Islands (marble on Paros and Naxos; emery, used for working the marble, on Naxos). The Early Bronze Age sees the proliferation of these figurines, and the development of several distinct forms (Figure 5.13). These figurines are found both on settlements and with burials, though the burial finds are the most common. The site of Dhaskaleio Kavas (Keros) is particularly interesting, in that a large mass of figurines, apparently broken deliberately, were found on the shoreline associated with stone vases and decorated pottery, presumably representing a ceremonial or ritual deposition (Dickinson 1994). Cycladic figurines are found outside the Cycladic Islands themselves, indicating contact between the islands and the Greek mainland, and between the Cyclades and Crete. A number of figurines were found, for example, at Aghios Kosmas, on the Saronic Gulf near Athens, in a cist grave cemetery which also produced Cycladic type pottery and stone vessels: it may not be necessary to see this site as a 'Cycladic colony', as postulated by Mylonas (1959), but it does indicate a close relationship between the communities of the Cycladic Islands and those of the Attic coast. Outside of Attica and Euboia, Cycladic figurines are rare in mainland Greece. Cycladic figurines are found in significant numbers in central Crete, as at Aghia Fotia (Davaras 1971) and Archanes (Sakellerakis 1977), where they are associated with burials. A local style of figurine (the Koumasa type) was made on Crete itself, probably in imitation of the Cycladic examples. Connections between the Cyclades, Crete and the Greek mainland are hardly surprising, given that the Cycladic Islands were an important source of raw materials used elsewhere in the Aegean: marble from Paros and Naxos, obsidian from Melos and silver from Siphnos (Gale 1980; Gale and Stos-Gale 1981). The circulation of Melian obsidian in particular was well established long before the beginning of the Bronze Age.

The Early Bronze Age evidence certainly does not suggest any pre-eminent role for Crete in the cultural development of the Aegean (though it is interesting to note that objects of hippo ivory and Egyptian stone bowls are present in small numbers in Early Minoan Crete, and not elsewhere in the Aegean (Krzyskowska 1988): this is important since it contrasts markedly with the evidence for the Middle Bronze Age. In most respects, particularly in terms of settlement structure and evolution, the processes which can be observed on Crete and the Cyclades are comparable to those observable on the Greek mainland. The Cycladic Islands do seem to have enjoyed a privileged position in relation to exchange and trade, probably by virtue of the raw materials which were extracted there, and because of their geographical position as stepping stones between Attica and Crete. The Cycladic marble

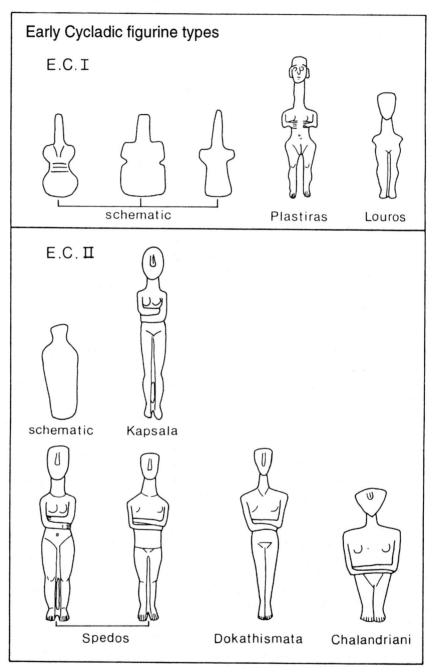

Figure 5.13 Early Cycladic figurine types

Source: Barber 1987

figurines also hint at the development of an insular religious tradition on the Aegean Islands, quite distinct from anything on the mainland (where figurines are found on the mainland, they are clearly Cycladic imports).

The Middle Bronze Age (*c.* 2200–1700 cal. BC)

The beginning of the Middle Minoan period (*c.* 2200–1900 cal. BC) saw the emergence of the first true 'palaces' in Crete, at Phaistos, Mallia, Knossos and Kato Zakros (Khania is also likely to have origins in this period, but the lower levels of this site have yet to be fully explored). These were impressive buildings with a cellular multi-room construction, including storage facilities (Figure 5.14). Where evidence is available, as at Mallia (Poursat 1987), it is clear that these palaces were associated with substantial towns. In many cases, understanding of this 'proto-palatial' period is hampered by the fact that the remains of the buildings underlie the constructions of later, more elaborate Bronze Age structures. Where the evidence has been recovered, however, it seems that these early palaces had paved western courts (this was certainly the case at Phaistos) and in some cases a central court (as at Kato Zakros): both are key features of the later palaces. To some extent the origins of these palaces can be traced back to cellular buiding complexes of the Early Minoan period, as at Vasiliki and Phournou Koryphe (Warren 1987). The Early Minoan II complex at Vasiliki, for example, incorporated a paved western court, whilst the contemporary complex at Phournou Koryphe included a 'storage magazine' with a number of pithoi in a small area. There are also specific architectural points of comparison between the Early Minoan complexes and those of the protopalatial period, such as the use of wooden beams within the wall construction at Vasiliki, and the existence of low benches around the walls at Phournou Koryphe (Warren 1987). It seems clear, then, that the palaces developed from local roots, albeit under a degree of Near Eastern influence, as suggested by Warren (*op. cit.*). This is also the period at which writing appears, initially in the form of 'hieroglyphic' inscriptions: comparison with later Linear A and Linear B tablets suggests a predominantly administrative function for these inscriptions. The distribution of 'hieroglyphic' and Linear A tablets is very clearly centred on Crete, but examples have also been found at Phylakopi, Aghia Irini, Akrotiri, Kastri (Kythera) and Aghios Stephanos, and a related script has been identified on Cyprus. Poursat (1987) points out that early inscriptions in Crete are not confined to the palaces: he identifies two 'administrative buildings' in Quartier Mu at Mallia, separate from the palace itself. These buildings have inscribed tablets and clay seals, together with evidence for storage and for craft production: some of the material found within these buildings (an unfinished stone kernos, clay moulds from triton shells and horns of agrimi) suggest a religious dimension. All of these are activities associated with the later palaces, and Poursat argues that these activities only became centralised and monopolised by the palaces at a later stage. It is clear, however, that by Middle Minoan IIIA (i.e. the developed 'First Palace period' *c.* 1800–1700 cal. BC), the palace at Knossos had acquired most of these functions (Cadogan 1987). A concentration of hieroglyphic tablets from the north end of the western corridor suggests that this part of the palace had already acquired a bureaucratic function, which continued into the Late Minoan period. Cadogan (*op. cit.*) also points to evidence for storage facilities in the early palace at Knossos, and suggests that the surplus agricultural produce stored in these facilities was used to support craft specialists. MacGillivray (1987) points out that the distribution of particular pottery styles in Crete mirrors that of the palaces and their putative territories, suggesting palatial control over ceramic production, despite the lack of evidence for pottery manufacture in the palace buildings themselves.

The storage facilities of the early palaces of Crete are clearly of great significance in attempting to understand the function of these buildings. Moody (1987) has stressed that the proportion of floor space dedicated to storage in the early palaces was, on average, significantly greater than in the later palaces. At Knossos, for example, in Middle Minoan II, storage takes up an estimated 1713 m^2 (12 per cent of the surface area): this figure declines to 1250 m^2 (9 per cent) in Middle Minoan III and to 850 m^2 (5 per cent) in Late Minoan I.

The early palaces of Crete must surely reflect the emergence of powerful local elites. Although Knossos was physically the largest of the palaces, there is no evidence to suggest that it had pre-eminent status over the other palaces during the First Palace period: it is more likely that we are dealing with 'peer polities'. The storage facilities suggest that these palaces were able to requisition a considerable agricultural surplus from their local communities, and this surplus was presumably used to support the elite themselves and the builders of the palaces, together with a variety of craft specialists. That this power structure was, at this time, unique to Crete, is demonstrated by the absence of comparable palace structures elsewhere in the Aegean. The second quarter of the second millennium cal. BC (Middle Cycladic period) was marked by the appearance of large nucleated settlements on many of the Cycladic Islands. Some of these, such as Phylakopi and Aghia Irini, can be considered as towns, but none have palatial centres. These islands, in any case, came increasingly under Cretan influence, as we shall see in the following chapter, and many of the cultural developments which can be identified in the Cyclades should perhaps be understood in the context of these influences.

If the palaces do reflect the emergence of powerful elite groupings, what was the basis of their power? Although the traditional model of Minoan 'thalassocracy' (cf. Hägg and Marinatos 1984) posits the control of trade as the central feature of the power base of the palatial elites, the main trade explosion post-dates the emergence of the palaces. Middle Minoan pottery is found in small quantities on Cycladic and Helladic sites, extending as far north as coastal Thessaly, and Cycladic pottery has also been found at Knossos

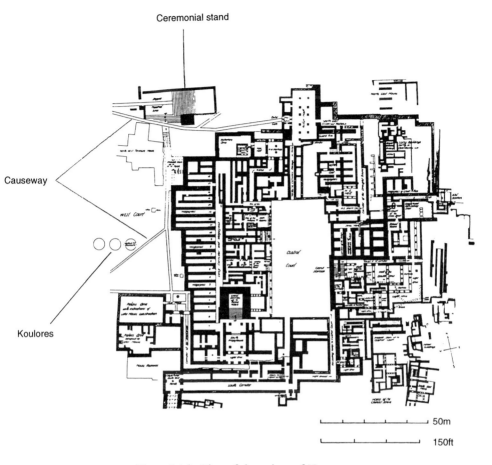

Ceremonial stand

Causeway

Koulores

50m

150ft

Figure 5.14 Plan of the palace of Knossos

Note: This is a composite plan, showing walls dating to several phases. The basic plan of the palace, however, including the Central and Western Courts, was established in the First Palace period

Source: Evans 1921

(MacGillivray 1984). A few sites, such as Lerna, have produced relatively large quantities of Cretan pottery, suggesting more organised directional trade. Crete also seems to have a near monopoly in the distribution of Near Eastern and Egyptian imports to the Aegean (Catling and MacGillivray 1983). Eighteenth-century BC documents from Mari in Syria mention 'Kaptaran' (convention-ally transalated as 'Cretan') merchants based at Ugarit trading in, amongst other things, weapons, textiles and pottery (Heltzer 1988, 1989). The evidence for trade and exchange will be considered in detail in the following chapter:

117

suffice to say at this stage that none of the evidence suggests that it was suffi-ciently important at an early stage for it to be considered as a 'prime mover' in the development of palace societies in Crete. Exchange, in any case, can only take place on the basis of the procurement of a significant surplus within the home community, which can be exchanged for other commodities. The evidence of the storage facilities in the Minoan early palaces suggests that this surplus was essentially an agricultural one, based on the 'Mediterranean poly-culture' of olive, vine and wheat (Renfrew 1972). To this list Halstead (1981) would add sheep and their products, and we might also consider the possible importance of fishing to an island society such as Crete.

There is an increasing body of evidence to suggest that the surplus which was stored in the earliest Minoan palaces was appropriated in the context of religious observances and obligations. Gesell (1987) identifies evidence for ritual and cult activities in the early palaces of Phaistos and Mallia. At Phaistos a shrine complex in rooms V–IX was distinguished by the presence of a fixed offering table and a variety of cult equipment (portable offering tables, stone bowls, a triton shell, etc.). In room VIII, annexes served as preparation rooms, in room V for grinding grain, and in rooms VI and VII for storage. The benches around the walls of room VIII itself would hold no more than twelve people, suggesting that participation in the ceremonies conducted there was restricted to a small number of people. Although this complex was originally accessible only from inside the palace, a door was later cut through to allow access from the west court (Gesell *op. cit.*). Directly to the north of this complex was a large open hearth, with a passage leading from the west court, which may have been used for public rituals, in contrast to the relatively private ceremonies conducted within room VIII. Gesell identifies three other groups of cult rooms in the early Phaistos palace, all associated directly with the west court. Again, these are relatively small: one (incorporating rooms LI,LIII–LVII and LXII) had evidence for food production, suggesting the preparation of a ritual meal, whilst another provided a quantity of small cult items, possibly used in the context of individual acts of preparation or worship. In both cases, these rooms were accessible from the public west court of the palace. A similar complex of rooms at Mallia, dating to the Middle Minoan II period, incorporated a sanctuary (with a fixed offering table), an ante-room and a storeroom. Again, cult equipment was found, including portable offering tables, a tripod base with double-axe decoration, a sherd with decoration in the form of 'horns of consecration', a clay model of a triton shell and two small animal figurines (Poursat 1966). Gesell (1987) estimates that around twenty-one people could have participated in cere-monies in this sanctuary at any one time. Sanctuaries of this type, however, are not unique to the palaces: Poursat (1987) identifies one in Quartier Mu (part of the town of Mallia), in a building which he identifies as having reli-gious and administrative functions. Another feature of the Minoan palaces which has been identified as having ceremonial functions is the so-called 'lustral basin' (Nordfeldt 1987). A lustral basin is a room with a shrunken floor approached by a staircase. The approach to the floor invariably involves at least one change in direction. Because of a superficial resemblance to small pools, these have often been interpreted as bathrooms (Graham 1959, 1977), but, as Nordfeldt (1987) points out, this is unlikely, since none have drains! Nordfeldt suggests a ritual function, perhaps as symbolic representations of cave sanctuaries. One such feature, in an obviously ritual context, is in the 'throne-room' complex at Knossos (Niemeier 1987). Although Evans (1921) considered the throne-room complex to have been added in Late Minoan II, it has recently been argued (Mirié 1979) that the complex, including the lustral basin, has its origin in the eighteenth century cal. BC.

The evidence for ceremonial or cult activities discussed above relates, for the main part, to small-scale and presumably private rituals, in which only a small number of people would have participated. The sanctuaries and lustral basins are small, and in many cases are only accessible from the inside of the palace. Whether the people who participated in such ceremonies were palace residents is open to question. Nordfeldt (1987), for example, sees no evidence for any residential quarters in the palaces, but it seems reasonable to suggest that they had privileged access to the interior. The west courts of the palaces are, by contrast, highly public areas, large, open, accessible and facing the surrounding town. These areas have been interpreted as arenas for public ritual (Marinatos 1987). The west courts at Phaistos and Knossos incorporate large circular pits, known as 'Koulores', which Marinatos interprets as granaries. In both cases there are causeways, entering the west court from the direction of the town: these causeways form a triangle in the west court itself, one branch of which leads to the koulores (Figure 5.15). This suggests the existence of ceremonies associated with the granaries, possibly, as Marinatos suggests, a celebration and presentation of the harvest. The 'causeways' are slightly raised above the level of the pavement, and are too narrow to have served for regular traffic, leading to Marinatos' suggestion that they served as 'processional roads'. Whilst at Knossos and Phaistos one branch leads to the koulores, another branch invariably leads to the western palace entrance (this is also true at Mallia). Marinatos points out that the rooms along the western facade of the palaces always incorporate storage facilities or 'magazines', and suggests that agricultural produce was ceremonially presented in the western court before being stored in these facilities. On the north side of the western court at Phaistos (Figure 5.15) is a stand consisting of nine steps leading to a narrow platform: the third branch of the causeway connects the palace entrance with this stand, suggesting that this feature played a central role in ceremonial activies in the west court. A similar arrangement is found at Knossos, and although this dates to the Second Palace period (Middle Minoan IIIA-Late Minoan I), Marinatos suggests that it may have replaced an earlier structure. She interprets these structures as ceremonial platforms for the officiants in west court ceremonies.

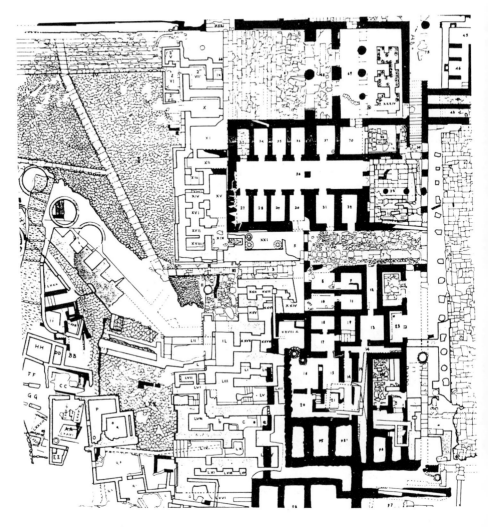

Figure 5.15 The western court of the First Palace at Phaistos, showing the position of causeways, koulores and ceremonial stand

Source: Marinatos 1987

It is clear, therefore, that the early palaces had a variety of ceremonial functions, involving both public and private ritual, and the evidence suggests that the public rituals may have included ceremonial presentations of grain and, perhaps, other commodities, by the populace to the palace elite. To gain a full understanding of religion in the Middle Minoan period, however, we need to look beyond the palaces themselves.

In addition to ritual arenas within the palaces, peak sanctuaries and cave sanctuaries are recorded (Faure 1963, 1965, 1967, 1969; Rutkowski 1986). The peak sanctuaries may have their origin in Early Minoan times (an arc-shaped structure above Phournou Koryphe dates to Early Minoan II, whilst the sanctuary at Petsofas seems to have been established in Early Minoan III), though most were established in Middle Minoan I, or later. Although Faure identifies around forty-five peak sanctuaries, certain sites stand out as being particularly important on account of their size and elaboration: he argues that each of the polities represented by the early palaces had one major sanctuary (Mount Juktas is presumed to have served Knossos, Mount Idha to have served Phaistos and Taostalos to have served Kato Zakros). This point is followed up by Cherry (1978), who considers peak sanctuaries as having played a fundamental role in the development of palace states (he argues that religious legitimation is more important in the establishment of social hierarchy than in its maintenance). The major peak sanctuaries are characterised by a temenos, or sacred enclosure, around the summit of the peak. Finds include offering tables, double axes, quantities of pottery and large numbers of animal and human figurines. Peatfield (1987) points out that, whereas the sanctuaries of the palaces mostly contain intact stored pots and cult objects, the assemblages from the peak sanctuaries are dominated by broken artefacts of similar types, suggesting the sanctuaries, rather than the palaces as the main arenas for ritual deposition. Many of the peak sanctuaries went out of use at the end of the First Palace period, but a few continued in use, and were greatly elaborated in the Second Palace period: these are the sanctuaries directly associated with the major palaces. Like the peak sanctuaries, the cave sanctuaries were foci for ritual deposition of objects (Rutkowski 1986). At Psycro, for example, the cave contained a lake, in which bronze figurines, sealstones, bronze rings and other items were deposited. Whilst much of the material from this, and other peak sanctuaries, is Late Minoan, some Early and Middle Minoan material was found in association with a small altar within the cave. Middle Minoan material has also been found in the cave sanctuary of Skotino. It has been suggested that the lustral basins of the Minoan palaces were symbolic representations of cave sanctuaries such as these.

Although the palaces themselves are found only in Crete, there are elements of Middle Minoan culture that are found elsewhere in the Aegean. Evidence for continued settlement nucleation can be identified on most of the Cycladic islands, with the growth of towns such as Phylakopi (Melos), Kastri (Kythera)

Aghia Irini (Keos) and Akrotiri (Thera). Some of these towns, such as Phylakopi and Kastri, have planned grid-type street patterns, similar to contemporary Cretan towns, such as Mochlos, Pseira and Gournia (Coldstream and Huxley 1984). Some, however, such as Aghia Irini, are also distinguished from Cretan towns by the presence of fortifications. Benzi (1984) and Branigan (1984) have argued that some of these settlements (specifically Kastri on Kythera and the Acropolis of Ialysos) should be seen as Cretan 'settlement colonies' (cf. Branigan 1981) with substantial populations of Minoan immigrants. Elsewhere in the Cyclades, as at Phylakopi (Barber 1984), Minoan imports are present, but in smaller numbers, and the bulk of the pottery in use was of an essentially local character. Palatial-type buildings, and the architectural features associated with them, are conspicuous only by their absence from the Cycladic Islands.

The 'First Palace' or 'Protopalatial' period lasted from the end of Middle Minoan I (*c.* 1900 cal. BC) to the end of Middle Minoan IIIA (*c.* 1700 cal. BC). At the end of this period the major palaces were destroyed, possibly by an earthquake. In the course of rebuilding the palaces, they were elaborated and extended: new types of feature appeared and the Minoan 'palace civilisation' reached its peak. The 'Second Palace period' (Middle Minoan IIIB to Late Minoan IB: *c.* 1700–1450 cal. BC) was also marked by a significant development of international trading contacts, with Crete at their centre, but this is essentially a matter for the following chapter.

The Late Bronze Age (*c.* 1700–1200 cal. BC)

The Late Bronze Age in Crete can be subdivided into two phases: the 'Second Palace period' (*c.* 1700–1450 cal. BC), during which the five major palaces (Knossos, Phaistos, Kato Zakros, Khania and Mallia) were rebuilt and developed, and the 'Third Palace period', in which only Knossos survived as a major palatial centre. The Second Palace period is also marked by the appearance of a new script, known as Linear A, and by a marked increase in the number of tablets found: these tablets, however, have not as yet been deciphered. Linear A tablets are concentrated in the major palaces, and in a very small number of other settlements, as at Aghia Triadha. The tablets of the Third Palace period are written in Linear B, which has been deciphered, and which clearly represents an early form of the Greek language.

The Second Palace period can be identified as a time of major settlement expansion. All of the major palaces were associated with towns, and that at Knossos appears to have covered a surface area of around 750,000 m², of which around 300,000 m² was covered by houses (Hood and Smyth 1981). Evidence for increased settlement density can also be identified away from the main palatial centres, and Warren (1984) has drawn attention to the increasing number of settlement sites in areas such as the Bay of Mirabello, the south coastal region and the upland plain of Lasithi (Watrous 1982).

Detailed study of the non-palatial settlements (Nixon 1987; McEnroe 1982) has revealed the presence of some buildings with palatial features, including ashlar masonry, lustral basins, storage areas and frescoes. These buildings are smaller than the major palaces, and are generally referred to as 'villas', though some should probably be seen as small palaces. A few, such as Aghia Triadha, have produced Linear A tablets, and many have produced luxury goods. Nixon (1987) distinguishes a class of 'non-typical outlying settlements', such as Gournia, Tylissos, Aghia Triadha and Kommos, characterised by the presence of at least two of these buildings. These settlements can be seen as sub-palatial centres, acting as intermediate points between palace and village for the redistribution of resources. McEnroe (1982) has identified a second category of large buildings in the outlying settlements, with non-ashlar construction, and generally without frescoes, luxury goods or Linear A tablets. A clear settlement hierarchy, therefore, seems to be emerging. Whilst Nixon (1987) interprets the rise of the outlying settlements as evidence for a decentralisation of power away from the palaces, this is not necessarily the case. The palaces, villas and outlying settlements seem to have functioned within an integrated system, the primary function of which was the mobilisation of an agricultural surplus towards the palaces, and a corresponding flow of small numbers of prestige items from the palaces back to the villas and other settlements (the manufacture of prestige goods seems to have taken place almost exclusively in the palatial centres). This can be seen as a classic example of a 'prestige goods economy', as outlined by Friedman and Rowlands (1977), with a patron–client relationship between the palace elites and the smaller-scale elites based in the villas. Some of the outlying settlements may have served a particular function within the palatial economy: Kommos, for example, seems to have served as a port for the palatial centre of Phaistos (Shaw 1987).

The Second Palace period is marked by the rebuilding of the major palaces, in a larger and more elaborate form than before. At Knossos, Phaistos, and Kato Zakros, the proportion of floor space devoted to storage was substantially reduced (Moody 1987). This decline in food storage seems to have been matched by an increasing emphasis on workshop and cult areas. Moody (*op. cit.*) notes that, at Knossos, many of the areas previously devoted to storage (rooms 41 and 42, the 'Lobby of the Stone Seat' and associated rooms, rooms adjoining the west court) were converted to these new uses, whilst at Phaistos, new areas dedicated to craft production were built to the northeast of the central court. There is clear evidence for manufacturing activities (particularly the manufacture of prestige goods) taking place within the palace complexes themselves. At Kato Zakros, for example, room 43 seems to have served as a workshop for marble objects, whilst room 44 was used for the manufacture of quartz pin-heads, and imported raw materials (notably elephant tusks and copper ingots) were stored on the upper floor of the west wing (Chrysolaki and Platon 1987). The existence of a steatite workshop next to the 'Treasury of the Shrine' at Kato Zakros could be taken to suggest a degree of theocratic control over the manufacturing activities (Chrysolaki and Platon *op. cit.*).

The rooms along the west side of the central court seem to have been developed as centres for cult activities in most of the later palaces (Gesell 1987). Dickinson (1994) emphasises the number of complexes of assumed ritual function which can exist in a single palace building, identifying seven such complexes in the second palace at Knossos (the central palace sanctuary, the throne-room complex, the east hall, the hall of the double axes, the royal villa, the house of the chancel screen and the high priest's house). Common features which distinguish these complexes include frescoes showing religious themes (there is a particular concentration of these at Knossos) and concentrations of cult objects, including offering tables, horns of consecration, double axes and figurines of a presumed female deity (usually depicted with a long skirt, bare breasts and often holding snakes).

As in the early palaces, two basic types of cult areas can be distinguished: small cult rooms, usually accessible only from the interior of the palaces, and larger areas for public cult, centred on the west courts of the palaces. Marinatos (1984a) stresses the link between the 'domestic sanctuaries' and food preparation, particularly the grinding of grain and the storage of produce, suggesting the preparation and consumption of ritual meals. At Mallia, for example, a small sanctuary in the south section of the palace, distinguished by the presence of an altar, clay feet and cult vessels, had annexes with evidence for food preparation and storage. The central courts of the palaces also seem to have been used for rituals involving small numbers of people: altars are present in the central courts at Phaistos and Mallia (Gesell 1987). Probably the best example of a 'private' cult area of the Second Palace period is the throne-room complex at Knossos (Figure 5.16). This complex, which includes a lustral basin, was established in the First Palace period (Mirié 1979), but was considerably altered in the Second Palace period (Niemeier 1987): stone benches were added along the walls of the throne-room itself (Figure 5.16(2)) and the adjacent antechamber (Figure 5.16(1)), and a stone seat (the throne) was placed against the north wall of the throne room, opposite the lustral basin (Figure 5.16(3)). Frescoes depicting griffins flank both the throne itself and the door between the throne room and the 'inner sanctuary' (Figure 5.16(4)). Niemeier (*op. cit.*) points to Minoan seals, showing the skirted and bare-breasted goddess flanked by griffins, and suggests that the throne room, far from being the reception room of a Minoan king, was in fact the scene of an 'epiphany' ritual, in which the goddess, personified by a priestess, entered the throne room from the inner sanctuary and was enthroned between the griffins of the fresco. She would also, presumably, conduct some form of ritual in the lustral basin. The rooms of the 'service section' (Figure 5.16 (5–7)), in the northwestern part of the complex, are interpreted by Niemeier as having served for the ritual preparation of the priestess, who would presumably have entered through the north door rather than through the 'antechamber'. Gesell (1987) estimates that twenty-three participants could have sat on the benches around the throne room itself, and that a further twenty-nine people could have stood in the room in front of the throne.

The west courts of the palaces were enlarged and made monumental as part of the rebuilding operation at the beginning of the Second Palace period (Marinatos 1987). The causeways at Phaistos were covered over, but those at Mallia and Knossos remained in place. At Knossos and Phaistos the koulores were filled in and paved over. At Mallia, however, the above-ground granaries remained in use, and the apex formed by the triangle of the causeways runs to their entrance, suggesting a continuation of the ceremonial practices established in the First Palace period. Hallager (1987) draws attention to the association, in the west wing at Knossos, of cult storage areas (the temple repositories), a sanctuary (the 'tripartite shrine') and storage areas, and suggests that this area may have been used for a form of 'harvest festival' celebration. The ceremonial platform at Phaistos went out of use, but a new one, of much the same type, was built in the west court at Knossos. Marinatos (1987) has attempted to reconstruct the rituals which took place in the western courts of the palaces, on the basis of the 'sacred grove and dance' fresco from Knossos. She argues convincingly that this fresco represents a ceremony in the west court, since the causeways are clearly shown. In the upper part of the scene, a large number of spectators are represented stylistically by a bank of heads. In the foreground, a group of women are dancing: they are dressed in long skirts, similar to those worn by depictions of the 'goddess' with the snakes. In the middle ground, on the causeways, are two groups of young men: one such group appears to be facing a single man, who is holding a staff, presumably an officer or leader. Saflund (1987) compares this depiction of the single male figure addressing a group of youths with the imagery of the 'chieftain cup' (Figure 5.17) from Aghia Triadha, which shows a male figure holding a staff, addressing a second male figure, who holds a baton. The staff and baton are closely comparable to those held respectively by the 'officer' and youths in the Knossos fresco, and there seems little doubt that we are dealing with a representation of a comparable ceremonial event. Saflund (*op. cit.*) interprets this in relation to Strabo's description (after Ephorus) of Cretan male initiation rites, in the tenth book of his *Geography*. Although the ceremonies described by Strabo are much later in date than those depicted on the Aghia Triadha cup or Knossos fresco, Saflund suggests that they may represent a continuation of Minoan ceremonial practice (this would, in itself be interesting, given the supposed 'collapse' of Minoan culture). According to Strabo, Cretan male youths were enrolled in 'herds' (Agelai), divided into age sets. Each Agela had a leader, known as the 'Bouagos', who was in turn subordinate to the 'Eiren', who had command over all of the Agelai. Saflund suggests that the individuals carrying staffs on

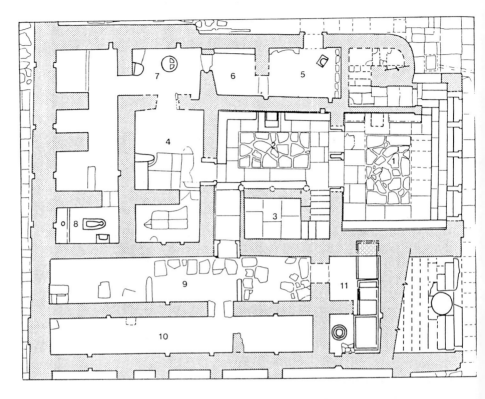

Figure 5.16   Plan of the 'throne room' complex at Knossos

Source: Neimeier 1987

both the Aghia Triadha cup and the Knossos frescoes represent Eirens, and that the second youth on the cup represents a Bouagos. She points out that the chieftain cup was found in an annex, to the southwest of the Aghia Triadha palace, and suggests that this may have served as accommodation for the Agelai during their period of training. Strabo also refers to collective marriage ceremonies for youths promoted from the Agelai, and Saflund suggests that the dancing women in the Knossos fresco may represent their prospective brides.

It seems likely, therefore, that the palaces of this period served as foci for at least three types of ritual:

1   Public rituals in the west court, involving the ceremonial presentation of the harvest to the palace: a continuation of a tradition established in the First Palace period.

126

2 Public rituals in the west court, involving male initiation ceremonies, and perhaps collective marriage.
3 Rituals involving far smaller numbers of people, held in the central court, and in the interior of the palaces, including the 'epiphany' ritual suggested for the Knossos throne-room complex.

These rituals may have been closely linked to one another, and may even have been integrated into a single complex ceremony or festival (it is interesting to note, for example, that the Agelai depicted in the Knossos fresco are small enough to have participated as a group in the 'epiphany' ritual in the throne room). If the palaces served as centres for communal rituals, it is less clear that they ever served as palaces! Nordfeldt (1987) questions whether there is any convincing evidence (other than the historical legend of King Minos) for residential quarters in the palaces, though there are plenty of rooms which *could* have served this role, and the palaces do have some features which might lead us to suspect this, including structures which have been interpreted as lavatories. All of the archaeological and iconographic evidence suggests that the palace rituals were controlled by an elite (there is restricted access to some of the ritual areas, and the frescoes show people of different status participating in rituals), and it is by no means unlikely that this elite was resident in the palaces. Whether there was a king is a different question entirely: the iconography shows both men and women presiding over ceremonies, but women seem to predominate. Given the reinterpretation of the throne room (Neimeier 1987), together with the evidence of the temple fresco from Knossos (Marinatos 1987), which shows women standing on a ceremonial stand similar to that identified in the west court, the idea of a priestess–queen is perhaps more convincing. As in the First Palace period, it is clear that religious activities were not confined to the palaces themselves. The votive deposition of material in the cave of Psycro, which began in the First Palace period, continued: material from the cave dating to Late Minoan I includes worshipper figurines, double-axe models, weapons, seals and jewellery (Rutkowski 1986). Whilst many of the peak sanctuaries went out of use at the end of the First Palace period, a few of these sites, such as Juktas, not only continued in use, but were greatly elaborated, with the addition of large cult buildings. These are the sanctuaries which Faure (1963,1965,1967,1969) considers to have been associated with the major palaces, and there is clear artefactual and iconographic evidence to link the cult activities of the sanctuaries with those of the palaces. The elaboration of this small number of peak sanctuaries, together with the mass of evidence for palace-based ritual, suggests an extension of palatial control over religion in the Second Palace period.

The Second Palace period was also marked by a dramatic explosion of trade and exchange in the eastern Mediterranean, in which Crete clearly played a major part. There is a significant growth of Minoan influence in

Figure 5.17   Detail of the 'chieftain cup' from Aghia Triadha

Source: Evans 1921

the Aegean area as a whole, marked by the proliferation of Minoan pottery and local imitations, particularly in the Cyclades (Davis 1986). For the first time, the developing centres of the Greek mainland became involved in this process of exchange with Crete: much of the material from the early shaft graves of Mycenae, for example, is of Cretan origin. Crete also seems to have served as a channel for the import of prestige materials from outside the Aegean world, including gold, ivory, ostrich eggs, Egyptian alabaster, lapis lazuli from Central Asia and copper and tin ingots from Cyprus and Anatolia. Caananite and Cypriot pottery is also found in Crete. It is significant that virtually all of the Aegean material found in Near Eastern (including Cypriot) and Egyptian contexts can be clearly identified as Cretan. The fact that a script related to Linear A (Cypro-Minoan) developed on Cyprus suggests a specific axis of trade between Crete and Cyprus. Communities on the Greek mainland seem to have been more concerned with developing contacts in the central Mediterranean world: material of Late Helladic I–II (*c.* 1550–1450 cal. BC) has been found in the Lipari Islands (Re 1986), southern Italy and Albania (Dickinson 1986; Knapp 1990). The involvement of Crete within wider networks of exchange and trade will be discussed in detail in the following chapter.

Crete, therefore, seems to have maintained its unique status within the Aegean world throughout the Second Palace period: the 'palace economy' of Crete remained unique (though mainland centres such as Mycenae were becoming increasingly wealthy and powerful), and Cretan communities seem to have been by far the most active players in trade and exchange, both within the Aegean region itself, and between the Aegean and areas to the East. The 'Minoanisation' of the Aegean, however, did see the spread of some aspects of Cretan culture, including religion, to other communities, particularly in the Cyclades. This process of acculturation is most evident from the spread of Minoan pottery, including both domestic pottery and ritual vessels such as rhyta. Occasional Linear A tablets are also found, as at Aghia Irini (Keos) and Phylakopi (Melos), though there are no archives outside Crete itself. Finds from a 'temple' at Aghia Irini are particularly revealing: the building itself is not of Minoan form, but the material from it includes fragments of at least fifty-five terracotta female figurines of classic Minoan type (with long skirts, bare breasts and snakes). On the nearby Troullos Hill were found two offering tables and a bronze figurine, in a context which calls to mind the Cretan peak sanctuaries (Davis 1984). Given, however, that the Aghia Irini temple is not a Minoan-type building, and that Troullos lacks the structures characteristic of the later peak sanctuaries in Crete, the overall impression given by these sites is of 'a Minoan veneer given to local cults' (Dickinson 1994) rather than a wholesale adoption of Minoan practices.

The evidence from the site of Akrotiri (Thera) is much less ambiguous (Marinatos 1984 a,b). At the beginning of Late Cycladic 1 (c. 1550 cal. BC), Akrotiri was little different from any other Cycladic town. Following destruction, however, probably by an earthquake, the town was rebuilt with much wider lanes, large open spaces and, together with blocks of houses of the normal Cycladic type, a series of individual houses (Figure 5.18: 'House of the Ladies', 'West House', Xestes 3 and 4), of clearly Minoan inspiration, closely comparable to the Cretan villas. Marinatos (1984a) has identified a number of shrines in the Late Cycladic town of Akrotiri: these are closely comparable to the shrines found in the Minoan palaces, and have been identified on the basis of the following criteria:

1 Iconographic content of frescoes.
2 Presence of cult equipment.
3 Architectural features.

The clearest evidence is that of the frescoes: a frieze in the shrine of the ladies, for example, appears to show a robing ritual, which calls to mind the epiphany ceremony postulated by Niemeier (1987) in the throne-room of Knossos. A fresco in Xeste 3 shows women gathering crocuses, and making offerings to a seated goddess figure, flanked by a griffin and a monkey.

It could be argued that the iconographic links evident in these two cases are not simply with Crete, but more specifically with Knossos. Cult equipment is present in large quantities in some of the shrines identified by Marinatos. In the 'mill of the square', for example, a shrine storeroom contained over 300 objects, including imported Minoan pottery, a hoard of stone vessels, triton shells, a gem with an engraved griffin and two ostrich egg rhyta. In the 'north mill', by contrast, the assemblage includes an offering table, several ceramic ewers and jugs, and a small bull rhyton: in this case, as Marinatos (1984a) stresses, the cult equipment is a meaningful set, intended for pouring libations onto the offering table. Significant architectural features in the Akrotiri shrines include a lustral basin in Xeste 3. Several of the shrines are associated with evidence for food preparation, another feature which links them to the shrines in the Cretan palaces. The shrine in the north mill, for example, forms part of a complex which also includes a grain mill, whilst the shrine of the ladies contained a cooking pot and a vat full of seeds. Pottery vessels in the shrine of the lillies were found to contain barley and bulbous plants, possibly onions or lilly bulbs. Marinatos (*op. cit.*) suggests that this shrine was used for a form of 'harvest festival': again, this is familiar from what we know of the Cretan palaces. The evidence from Akrotiri is quite unlike anything else found in the Aegean outside Crete: the Minoan character of the town and its buildings is so marked that it has even been suggested that we are dealing with some form of Cretan colony. Even here, however, we have (as far as is known at this stage) no palace, and much of the pottery is locally made. It seems likely that we are dealing with an essentially Cycladic town (Doumas 1982), with a community of wealthy merchants (Schachermeyer 1980), perhaps including a small number of Cretans. The Cretan influence could even be seen simply as the result of exchange and trade, and the operation of the 'Versailles effect' (Wiener 1984), whereby local elites compete for prestige by emulating the architecture and associated cultural practices of a wealthier and more developed centre.

Between Middle Minoan I (*c.* 2200–1800 cal. BC) and Late Minoan IB (*c.* 1500–1450 cal. BC), a classic core–periphery situation seems to have developed in the Aegean world, with the core very clearly situated in Crete. The existence of the palaces suggests that society in Crete was more hierarchical than elsewhere in the Aegean, and the specific nature of the evidence from the palaces suggests that the power structures of which they were part operated very largely through the control of elaborate ritual practices. Marinatos (1984a) uses the term 'threskeiocracy' to describe the extent to which sacred and secular power were intertwined in Minoan society. The extent of hierarchy seems to have been greater in the Second Palace period than in the First, with the appearance of new social strata, represented by the villas. Similarly, we can trace a process whereby particular centres, most notably Knossos, became increasingly important, suggesting a gradual trend towards the development of hegemony. In the First Palace period, the effect of these developments on the rest of the Aegean

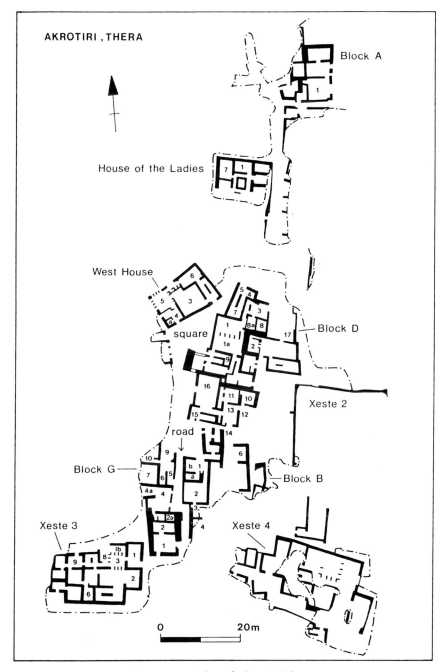

Figure 5.18 Plan of Akrotiri, Thera

Source: Marthari 1984

world appear to have been relatively limited: in this respect the early palaces can be compared to the Maltese temples, reflecting a specific insular elaboration of particular ritual practices. In the Second Palace period, however, Cretan communities acquired an influence in the eastern Mediterranean that was out of all proportion to the island's size and presumed military power. This seems to have been a result both of the Versailles effect, and of the increasing domination by Crete of networks of trade and exchange linking the Aegean to the Near East and Egypt. The notion of a Minoan thalassocracy depends upon this idea of King Minos and his dynasty controlling the seaways, but it is important to realise that this situation only really developed in the Second Palace period. Cretan society was far more isolated in the First Palace period, when the 'palace economy' developed. Thalassocracy, therefore, was secondary to threskeiocracy. The palace civilisation of Crete began as just another insular elaboration: unlike the other examples which we have looked at, however, it developed into something far more important and far more influential. We will explore some possible reasons for this at the end of this chapter.

All of the Cretan palaces, apart from Knossos, were destroyed at the end of Late Minoan IB, marking the beginning of the Third Palace period. Villas also disappeared as a class. Most of the palatial and sub-palatial buildings were permanently abandoned, though some of the palaces, such as Khania, were reoccupied in the Third Palace period (Moody 1983; Luckerman and Moody 1985). Three very different explanations have been proposed for this sequence of events. The settlement of Akrotiri (Thera) was destroyed by a volcanic eruption, and Marinatos (1939) considered that this eruption might also have been responsible for the destruction of the Minoan palaces. This explanation is now generally rejected on chronological grounds (Niemeier 1980), as the Thera eruption seems to have occurred significantly earlier than the destruction of the Cretan palaces. The explanation that has gained the most general acceptance is that of an invasion of Crete by Myceneans from the Greek mainland, and it is certainly true that the Third Palace period is marked by a dramatic decline in Minoan influence in the Aegean, and by a corresponding increase in Mycenean influence. At Knossos itself, Linear A is replaced by Linear B, a mainland script which, in linguistic terms, is recognisably Greek. Iakovidis (1979) argues that the Mycenean expansion into the Aegean began in the Second Palace period, citing iconographic evidence from Akrotiri. This idea is developed by Laffineur (1984), though it is challenged, for example, by Doumas (in Hägg and Marinatos 1984, 139). Whilst it is difficult to deny that Knossos was ultimately taken over by a Mycenean elite, Niemeier (1984) argues that this only happened after the fall of the palaces. He points out that the destruction horizons of the various Cretan palaces, sub-palatial settlements and villas are not contemporary, and cannot therefore be interpreted as resulting from a single event, such as a Mycenean invasion. At Phaistos, Mallia and Palaiokastro, for example, the destruction horizon dates to Late Minoan IB (c. 1450 cal. BC), whereas at Nirou Khania and Tylissos it dates to Late Minoan II. Pottery of Late Minoan II type can also be identified in destruction horizons from sites on Keos, Melos and Kythera, suggesting that the upheavals of this period were not confined to Crete. Niemeier (1984) also points out that the Linear B tablets from Knossos date to Late Minoan IIIB, and that the 'palace-style' pottery, in use at Knossos throughout Late Minoan II, is clearly of Minoan, rather than Mycenaean type. He suggests that the destruction of the palaces relates to a long period of troubles on Crete, probably involving wars between the elite based at Knossos and those centred on the other palaces: according to this model, the Mycenaean attack took place in Late Minoan III, taking advantage of the chaos on Crete brought about by internecine conflicts. Hiller (1984), similarly, argues for a relatively late Mycenaean takeover of Crete, not earlier than *c*. 1380 cal. BC. This is the period which sees the appearance of megaron buildings in Crete, as at Aghia Triadha and Khania Kydonia. These are large open halls with a central hearth, typical of Mycenaean high-status settlements. Further indications of Mycenaean incursion are provided by a shaft grave enclosure at Archanes, and a fortified acropolis at Kastrokephala Almyrou, and by the increasingly Mycenaean character of material culture assemblages in Late Minoan III. Mycenaean features can also be identified on the Cycladic Islands: large quantities of mainland pottery are found on Cycladic sites, and a megaron building has been identified at Phylakopi (Melos). The distribution of Mycenaean material within the east Mediterranean area suggests that communities on the Greek mainland also took over the exchange networks linking the Aegean to the Near East and Egypt.

At Knossos itself, the Third Palace period is marked by considerable evidence for continuity, and the palace, including the throne room, continued to be modified (Mirié 1979). Ritual practices continued within a broadly Minoan tradition, as shown by the establishment of the 'sanctuary of the double axes' in Late Minoan IIIA1. If the new overlords were Mycenaeans they had, to some extent 'gone native', appropriating elements of Minoan culture as part of their own strategy of legitimation. In most respects, however, Crete had, by the beginning of Late Minoan III, become integrated within a Helladic cultural complex centred on the Greek mainland – the core–periphery situation which had prevailed in the Aegean since the beginning of the Middle Bronze Age had been effectively reversed. Whilst communities in Crete itself continued to produce pottery in local styles until the end of the Bronze Age, those elsewhere in the Aegean adopted a relatively uniform Mycenaean material culture package. The major exceptions to this generalisation are the islands of Karpathos, Saros and Kasos (Melas 1985), where the material culture remains predominantly Minoan in character until the end of the Bronze Age. In the transitional period between Late Minoan IIIA2 and IIIB, imported and locally made Minoan pottery accounts for 64.3 per cent of the recorded pottery from these islands, whilst locally made and imported Mycenaean pottery accounts for 21.4 per cent and 14.3 per cent, respectively. In Late Minoan IIIB and Late Helladic IIIB1, however, locally made Minoan pottery accounts for 66.7 per cent of the assemblage, with Cretan imports representing 25 per cent (Melas *op. cit.*). To some extent this can be explained by the island's close proximity to Crete itself, but the remarkably limited extent of Mycenean influence must surely be seen as an example of insular conservatism.

UNDERSTANDING CULTURAL ELABORATION IN THE MEDITERRANEAN ISLANDS

In this chapter we have explored Evans' (1973) notion of cultural elaboration on islands. The most obvious common thread which runs through all of the case-studies which we have examined is that these island communities all displayed a remarkable level of cultural uniqueness: the culture of these islands, at these particular points in time, was quite distinctly different from that of adjacent mainlands and neighbouring islands. This is not true of all Mediterranean islands. The culture of Sicily, for example, was always closely linked to that of southern Italy, whilst that of the Ionian Islands was, in almost all respects, indistinguishable from that of the eastern Greek mainland. Similarly, if one looks at Corsica and Sardinia in the Neolithic, at Malta in the pre-Temple phase, or at Crete in the Iron Age, cultural uniqueness is far less marked than in the periods examined above.

Cultural uniqueness in island communities can be expressed in a number of ways: through material culture, religious practice and iconography, architecture and social organisation. It may be either innovative (as in the case of the Maltese temples or the Minoan palaces) or conservative (as in the case of the Late Bronze Age of Karpathos). In the cases examined above, cultural uniqueness appears not to have been an accidental result of isolation from the main stream of regional development, but rather the deliberate emphasising of an island community's unique cultural identity. This emphasis is likely to have been significant in the establishment and mediation of social relations both within island communities, and between those communities and their neighbours on mainlands and other islands.

What is most significant about the case studies which we have examined, however, is not simply the expression of cultural uniqueness but the scale of cultural elaboration. The taulas and talayots of the Balearic Islands, the nuraghi of Sardinia, the torri of Corsica, the stone temples of Malta and the palaces of Minoan Crete all represent significant investments of labour by island societies. This is the phenomenon noted by Evans (1973) as 'exaggerated development', and by Sahlins (1955) as 'esoteric efflorescence' in island societies.

In most cases, this exaggerated development relates exclusively to one aspect of the cultural life of an island community, very often a religious aspect. It would be difficult to see the Balearic taulas or the Maltese temples, for example, as anything other than religious monuments, and recent work (cf. Marinatos 1984a, 1987) has highlighted the religious dimension of the Minoan palaces. The precise function of the Balearic talayots, the Sardinian nuraghi and the Corsican torri is more enigmatic, though the evidence from the torre of Filitosa suggests that these monuments may also have had a religious dimension. In the terms identified by Stoddart et al. (1993), the communities which we have looked at can be described as 'monument oriented', meaning that surplus production was channelled primarily into the construction of monuments, and into activities (ceremonies, feasting, etc.) associated with them. The construction and use of monuments is likely, therefore, to have been of primary importance for the establishment, negotiation and transformation of social relations within these societies. Some indications of the possible sociocultural significance of monuments can be obtained from an examination of ethnographic and ethnohistorical case-studies (see Chapter 2), such as the ahus of Easter Island (Métraux 1957; McCall 1979) and the Nimangki/Maki monuments of Malekula (Deacon 1934; Layard 1942). In the case of Malekula the construction of monuments took place in the context of male initiation rites and, as such, played a key role in the establishment and maintenance of asymmetric social relations between men and women, and between men of different age sets. The Easter Island monuments were built in the context of a much more hierarchical social system, in which asymmetric relations were based not only upon gender and age, but also upon class (Métraux op. cit. identifies six distinct social classes in Easter Island society): these monuments, unlike those of Malekula, also seem to have served as 'territorial markers' for individual clan groups (McCall 1979). There is also a massive difference of scale, in that the ahus of Easter Island are very much larger than the megaliths of Malekula (similar differences of scale could be identified between, say, the Balearic taulas and the Minoan palaces). The monuments of Malekula and Easter Island functioned within very different social, cultural and symbolic contexts: the similarity between these two case studies lies in the centrality of monumental ritual to the social reproduction of these two very different communities. Both Malekula and Easter Island can, in this sense, be seen as 'monument-oriented' societies, and both conform to Meillassoux's (1972) model, in which asymmetric social relations are established and affirmed through elite control of sacred knowledge and ritual practice. None of this, of course, is unique to island communities: the megalithic traditions of Iberia (Leisner and Leisner 1943) and Brittany (Patton 1993) are classic examples of monument-oriented societies in mainland contexts. Such traditions, however, do appear both to be more common and more intense in island communities, and it is surely significant, for example, that the only prehistoric communities in the central and eastern Mediterranean which could be thought of as monument-oriented were on islands. A number of models have been proposed to explain this phenomenon, and these can be summarised as follows.

135

1 The *ecological model* focuses on the limitations of the island ecosystem (see Chapter 3), and seeks to explain monumental traditions as an adaptation to an environment of scarcity. Sahlins (1955) considers that the ecological constraints of island life prevent the direction of communal labour into substantive improvements in subsistence production, and that efforts are consequently channelled into the 'esoteric domain of culture' (the 'imaginary conditions of production' as discussed by Friedman and Rowlands (1977)). Following Renfrew's (1976) concept of monuments as 'territorial markers', we could equally argue that increased competition for land on a small island would lead to increased investment of land in monuments.

2 The *cultural isolation model*, by contrast, seeks to explain cultural divergence in island communities in much the same way as Darwin explained genetic divergence in island species. Vayda and Rappaport (1963), for example, argue for a cultural version of the genetic 'founder effect' identified by Mayr (1954): just as a founder population of animals will carry with it only a small proportion of the genes available in the parent population, so a human founder population will carry with it only a small proportion of the cultural characteristics, which are then exaggerated and accentuated through isolated development. Evans (1973) suggests that it is the relative security of island life that allows the elaboration of such cultural trends which would otherwise be arrested by extraneous factors.

3 The *sociogeographic model* focuses on social structure, and on the role which monuments play in the establishment and transformation of social relations. It can be argued (cf. Meillassoux 1972; Bender 1985) that, in most mainland societies, power relations are based partly on control of sacred knowledge and ritual practice, and partly on control over the circulation of socially valued material items (Patton 1993). Islands, being bounded and, to a greater or lesser extent, isolated, may give rise to the exclusive intensification of one of these phenomena: thus Stoddart *et al.* (1993) identify a cycle of monument-oriented and exchange-oriented societies in Neolithic Malta.

If there is a specific ecological, cultural or sociogeographic explanation for the kind of phenomena which we have been examining, one might expect cultural elaboration to be affected by particular geographical variables, such as island size and distance from mainland. The clearest examples of insular cultural elaboration in the prehistory of the Mediterranean are to be found in the Balearic Islands, Sardinia, Corsica, Malta and Crete. The biogeographic characteristics of these islands are summarised on Table 5.1.

There is clearly a correlation between degree of remoteness and evidence for cultural elaboration. Eighty-seven per cent of Mediterranean islands can be reached from the mainland without undertaking any single sea crossing of more than 40 km: all the islands listed above, therefore, fall within the top 13 per cent

Table 5.1 Biogeographic characteristics of islands with evidence for cultural elaboration

Island	Distance[1] (km)	Area (km²)	Biogeographic ranking[2]
Balearics	92	3740 (Mallorca)	40.7
		702 (Menorca)	7.6
Sardinia	58	24089	415.3
Corsica	58	8722	150.4
Malta	80	237	3.0
Crete	48	8259	172.1

Note: 1 Defined as the longest single sea-crossing required to reach a given island from the mainland. 2 Defined as distance/area (see Chapter 3)

bracket representing the most remote islands in the Mediterranean. There is also a correlation with island size: all of the islands listed above fall into the 24 per cent of Mediterranean islands with surface areas greater than 200 km². The islands with the greatest degree of cultural elaboration, therefore, are those which are both relatively large, and relatively distant from the mainland.

The ecological model would lead us to expect the greatest degree of cultural elaboration on those islands which are most environmentally circumscribed: in other words, those islands with the lowest biogeographic rankings. In fact none of the islands listed above fall within the 48 per cent of Mediterranean islands with biogeographic rankings less than 3, whilst three of these islands (Sardinia, Corsica and Crete) fall within the 10 per cent of Mediterranean islands with biogeographic rankings greater than 100. The reason for this is that, whilst there is a positive correlation between cultural elaboration and distance, there is also a positive correlation with island size: cultural elaboration is found primarily on the larger islands rather than the smaller ones. It is possible to conceive of environmental circumscription being a factor on Malta, or perhaps even on Menorca, but very difficult to imagine this being the case on islands as large as Sardinia, Corsica, Crete or Mallorca.

The cultural isolation model would lead us to expect a positive correlation between cultural elaboration and remoteness, which indeed we do find, but it would hardly explain the correlation with island size. The islands which would experience the greatest degree of cultural isolation would presumably be those which were visited least often by outsiders: we would expect these to be the smaller remote islands, rather than the larger ones. In any case, the archaeological evidence certainly does not suggest that the Balearic Islands were culturally isolated during the Talayotic period, or that Crete was ever truly isolated during the Bronze Age.

The evidence, on the whole, favours the sociogeographic model as the best explanation for cultural elaboration in the prehistory of the Mediterranean, though this is not to deny that factors of ecology and cultural isolation were significant in particular cases: both may have been important, for example, in

Malta, which is both small and remote. The islands which have produced the clearest indications of cultural elaboration are (as one would expect on the basis of the sociogeographic model) those that are both large enough to support significant populations and generate an agricultural surplus for investment in communal projects, and sufficiently remote to exclude casual and unrestricted contact with mainland communities. The communities on these islands were, at particular moments in time, connected to communities on mainlands and other islands by complex systems of exchange and trade which seem to have had considerable social significance (see Chapter 6). When these exchange systems collapsed, however, due either to internal or to external socioeconomic factors, the effort which had previously been channelled into such contacts was redirected into monumentalism. It is even possible that control of monumentalism and control of exchange were, under certain circumstances, employed as rival strategies by competing elite groups, and that the transition from a 'monument-oriented' to an 'exchange-oriented' system may, in these cases, reflect the replacement of one elite structure by another. These are themes to which we shall return in Chapter 7.

6

NETWORKS OF
INTERACTION IN
MEDITERRANEAN ISLAND
PREHISTORY

Unlike the islands of the Pacific, none of the Mediterranean islands are truly isolated: the vast majority are within sight of either an adjacent mainland, or of other islands which serve as 'stepping stones'. It is hardly surprising, therefore, that many of the prehistoric communities on the Mediterranean islands were engaged in trade and exchange both with other islands and with mainland communities. In some instances, these exchange relationships can be explained in relatively simple economic terms: island communities either had resources which were required elsewhere (e.g. obsidian on Melos) or required resources which were not locally available (e.g. bronze on Malta or the Balearic Islands). In other instances, however, inter-island and island–mainland relationships appear to have taken on a far greater social significance.

In her book *Ulysses' Sail*, Mary Helms (1988) explores the 'widespread association of political elites with foreign and distant goods and information'. She argues that space and distance are commonly accorded political and ideological significance, and that geographical distance from a given point may be equated with supernatural distance. Islands, she suggests, with their natural zones of dry land, littoral and sea, offer obvious material for this kind of 'cosmological zoning'. The ethnographic literature, particularly in the Pacific region, contains numerous examples of this. Tambiah (1983), for example, in looking at Trobriand cosmology, describes myths and fantasies in which canoe travels as part of the Kula exchange network are compared with magical trips through the skies, whilst Williamson (1933) discusses the widespread Polynesian belief that the souls of the dead migrate across the sea to the horizon. Helms (*op. cit.*) suggests that the long-distance traveller may be considered to be the personification of a mythical hero, as in Trobriand Kula, according to Montague (1980), and among the Elema of Eastern Papua New Guinea, according to Williams (1977). Both Munn (1983) and Uberoi (1971) point to the way in which travel is often ritualised. The Kula exchange network of the Melanesian Massim (Malinowski 1922; Leach and Leach 1983; Munn 1986; Weiner 1976) is a classic example of an inter-island and island–mainland exchange system which plays a fundamental role in the negotiation and articulation of social relationships within the communities concerned (see Chapter 2). Kula plays an important role in the maintenance of asymmetric relationships between men and women (since in most communities only men engage in Kula), between elders and younger men (since young men can only enter the exchange network with the assistance of elders), and between commoners and chiefs (since the canoes used for Kula voyages are controlled by the latter). Although, as Helms (1988) demonstrates, long-distance travel and exchange may be socially significant in mainland societies, islands offer a greater potential for this, since contact with the outside world necessarily involves a sea voyage, and access to boats, trading partners and navigational skills can all potentially be controlled and monopolised by an elite.

In the previous chapter, following Stoddart *et al.* (1993) we identified a distinction between 'monument-oriented' and 'exchange-oriented' societies, and suggested a cyclical relationship in some instances (e.g. the Neolithic of Malta), with exchange-oriented social systems being replaced by those focused on monuments and vice versa. Whilst Chapter 5 was primarily devoted to monument-oriented societies, this chapter will look at the archaeological evidence for exchange-oriented societies in the prehistory of the Mediterranean islands.

There are several problems which must be faced in assessing the significance of exchange and interaction within the prehistoric communities of the Mediterranean islands. First, the archaeological evidence includes only those exchanged commodities which are durable, and these may represent only the tip of a much larger iceberg. In ethnographic case studies, socially significant exchange networks have involved the circulation of textiles and elaborately carved wooden objects, for example, which would not have survived in the soil conditions of the Mediterranean islands. Second, only a small proportion of those items found in the archaeological record can be reliably characterised, so that it is often not possible to establish whether an object was made locally or imported from elsewhere. Finally, the identification of exchange-oriented societies, as suggested by Stoddart *et al.* (1993), refers not to the presence or absence of exchange, but rather to the significance of this exchange within the social formation of an island community: inevitably, this is far more difficult to assess, and depends largely on the nature of the exchanged items and the contexts in which these items are found.

As in the previous chapter, the discussion in this chapter will be based around a series of specific case studies, which illustrate more general points.

ISLAND AND MAINLAND IN THE AEGEAN NEOLITHIC

The earliest documented examples of island–mainland exchange in the Mediterranean are in the Aegean. Obsidian from the Cycladic island of Melos

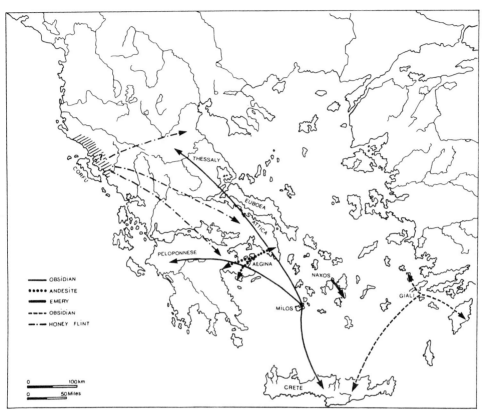

Figure 6.1 Exchange systems in the Aegean Neolithic

Source: Perlès 1992

has been found at the Francthi Cave, in mainland Greece, in Early Mesolithic contexts (Perlès 1979). There is, however, no evidence for any settlement on Melos itself prior to the fifth millennium cal. BC (Cherry 1981), so we are presumably dealing with direct access rather than exchange *strictu sensu*, probably procurement of obsidian by mainland communities in the course of long-distance fishing expeditions (Bintliff 1977; Perlès 1979). The distribution of Melian obsidian in the Neolithic covers the whole of the Aegean and most of mainland Greece (Renfrew *et al.* 1965; Torrence 1986). Virtually all of the obsidian used in the Aegean is derived from two outcrops – Sta Nychia and Demenegaki on Melos.

Although most studies to date have concentrated on the exchange of specific materials (most notably obsidian), Perlès (1992) stresses the need to place

141

this evidence in a broader context, and to consider all materials in circulation within an exchange system. She identifies three sharply differentiated systems of production and distribution involving different types of goods (stone assemblages, pottery and ornaments) and organised along fundamentally different lines.

Melian obsidian is by far the most abundant item of exchange in the Aegean Neolithic, being found on virtually all Neolithic sites within the region (Perlès 1992). Torrence (1986) has undertaken a detailed study of the evidence for obsidian procurement on Melos. Her surveys of the obsidian sources of Demenegaki and Sta Nychia permitted the following estimates of the total output of these sources in terms of obsidian macrocores: Sta Nychia, 4,895,870 (800 tonnes); Demenegaki, 3,084,338 (500 tonnes). Whilst these figures initially seem impressive, it should be borne in mind that the exploitation of obsidian on Melos took place over a period of at least 3000 years, leading to an estimated average production of only 2660 macrocores per year. This would correspond to approximately eighty-two labour days per year in the case of Sta Nychia, and fifty-one in the case of Demenegaki. Whilst this could correspond to a single part-time specialist operating at both quarries during the summer months, it could equally relate to many visits by small parties of people. The production of 2660 cores per year is also unlikely to have satisfied demand in the Aegean area, and it therefore seems likely that much of the obsidian left the quarries in the form of unmodified nodules.

There is, in any case, no archaeological evidence to suggest that Melos was permanently settled during the Early Neolithic period, when its obsidian resources were first exploited on a significant scale, and Torrence (1986) favours Bintliff's (1977) suggestion that obsidian was initially procured in the context of long-distance fishing expeditions.

Renfrew (1977) develops the 'law of monotonic decrement', according to which: 'In circumstances of uniform loss or deposition, and in the absence of highly organised directional exchange, the curve of frequency or abundance of occurrence . . . against distance from a localised source will be a monotonic decreasing one.' This law operates, quite simply, because as one moves further away from the source of a commodity, the cost of procuring that commodity increases. Torrence (1986) points out that Neolithic obsidian debitage in the Aegean actually increases in size with distance from Melos, in direct contradiction to Renfrew's law. This is in contrast to the Bronze Age evidence, where the law of monotonic decrement seems to apply. Torrence argues that if obsidian in the Neolithic was procured in the context of fishing expeditions, no additional cost would be incurred, since: 'Collection of raw material was a function not of consumer need, nor of distance from source to settlement, but was controlled by the fishing strategy.' On balance, therefore, the archaeological evidence seems to favour a 'direct access' model for the procurement of obsidian on Melos during the Neolithic. This does not necessarily mean, however, that all mainland communities obtained obsidian by sailing directly to Melos. Perlès (1992) suggests that both obsidian procurement and the manufacture of narrow blades were undertaken by a relatively small group of specialists, pointing out that obsidian implements in Neolithic Greece seem to have been made with a greater degree of skill than implements of locally available materials, such as chert. The distribution of Melian obsidian shows very little fall-off between south Greece and Thessaly, but a sudden fall-off between Thessaly and western Macedonia: this pattern of distribution corresponds closely with that expected from a system of 'middleman trade' (Renfrew 1984). Perlès (1992) goes on from this to develop a model for obsidian exchange based on itinerant workers, procuring obsidian on Melos and producing blades to order within local communities in Greece. Such workers, however, would not necessarily have been full-time obsidian specialists, and it is not difficult to see how this role might have been combined with the long-distance fishing activities envisaged by Bintliff (1977) and Torrence (1986). Van Andel and Runnels (1988) take the argument a stage further, suggesting that some coastal sites, such as Francthi, served as key points or 'emporia' for the exchange of obsidian and other commodities. It is conceivable that these sites, which do seem to be characterised by particular concentrations of imported material, were the home bases of maritime communities which specialised in long-distance fishing and exchange, and it would not be surprising if such communities also developed particular skills in obsidian working.

Obsidian was not the only lithic resource circulating within the Aegean Neolithic. Andesite, a hard volcanic rock from the island of Aegina, was used to make querns and grinding stones, which are found on the Greek mainland, though with a much more restricted distribution than Melian obsidian (Figure 6.1). The quantities of andesite imported to the Greek mainland seem to have been relatively small (only 10 to 30 per cent of grinding stones found within the area of Attica and the Argolid), but once again the mainland site of Francthi has a significant concentration of imported material, with around twenty objects of Aeginitan andesite (Perlès 1992). Emery from the island of Naxos was another exchange item, and is found on a number of other Cycladic islands, where it was used in the manufacture of ground stone axes.

These lithic resources from the Aegean islands appear to have been exploited primarily by mainland communities, at a time when settlement on the islands themselves was extremely limited. Mineral resources (obsidian, andesite, emery) and animal resources (fish, perhaps especially tuna) were probably procured by these communities as part of a single exploitation strategy, and it is likely that the existence of these resources was a significant factor in encouraging the settlement of the Aegean islands during the Neolithic and Early Bronze Age. Lithic assemblages seem to have been essentially utilitarian in character (Perlès 1992), and Melian obsidian is found on virtually all Neolithic settlements in the Cyclades, southern Greece and

Thessaly, suggesting that all communities had access to it. There is no evidence for inter-site stylistic variation.

Pottery appears to have been produced at a much more localised (probably domestic) level, as suggested both by stylistic and petrographic evidence (Perlès 1992). Stylistic variation is much greater, and Perlès suggests that this may have been significant in differentiating exchanges between partners and relatives in close social relations, sharing common symbolic codes, and more neutral exchanges between individuals with minimal social relations. The absence of external carbon sooting and internal residues on many pottery vessels also hints at a non-utilitarian function (Perlès *op. cit.*). Clear regional styles can be distinguished, suggesting small-scale regional circulation.

There is a third category of exchanged items in the Aegean Neolithic, comprising jewellery, stone vases and figurines. These are found over large areas, but always in small quantities, suggesting that access was more restricted (stone vases, for example, are recorded from only ten sites in Greece). Yet again, the evidence suggests that certain sites, notably Francthi, had a special role. Francthi has been identified as an 'export workshop' for the production of beads made from the shell of a flat cockle, *Cerastoderma glauca*: hundreds of blanks for these beads were found in the Francthi Cave, together with a specialised flint assemblage used for their production (Perlès 1992). The context of deposition of stone vases, jewellery and figurines is, in some cases, suggestive of a ritual significance: at Francthi, for example, a small marble vase (the marble imported from the Cyclades) was found associated with a child burial. The exchange of items of this nature is likely to have had considerable significance in relation to the establishment and creation of asymmetric social relations within and between communities.

Perlès (1992) argues convincingly that the production and exchange of lithic assemblages, ceramics and prestige goods (jewellery, stone vases, figurines) were organised in the context of fundamentally different social relationships. Lithic assemblages appear to have been essentially utilitarian in function: there is evidence for craft specialisation and for large-scale distribution in terms of both quantity and distance. Pottery, on the other hand, appears in many cases to have been non-utilitarian, shows a far greater degree of stylistic variation and circulated on a much smaller scale. Perlès (*op. cit.*) suggests that its circulation was important in the establishment of alliances and social relations at a local level. Prestige goods, like lithic assemblages, show evidence for specialised production, but were produced only in small quantities. Like lithic assemblages, however, they circulated over long distances.

There are, in fact, a number of points of similarity between the exchange of lithic assemblages and that of prestige goods. Both involved the circulation of raw materials derived from the coast (shell beads) and, more specifically, from the Cycladic Islands (Melian obsidian, Aeginitan andesite, Naxian emery, vases of Cycladic marble). The distribution of Melian obsidian, marble vases and *Cerastoderma* beads covers a similar geographical area (southern and central Greece) to that of Melian obsidian, though the obsidian is found in much larger quantities. Finally, certain sites, such as Francthi, seem to have played a central role in the production and circulation of both utilitarian lithic assemblages and prestige goods. It does seem likely that the procurement and circulation of obsidian was controlled, in the initial stages, by relatively small mainland groups, who had become acquainted with the Aegean islands in the context of long-distance fishing expeditions. These same fishing expeditions are likely to have brought these groups into contact with communities along the southern and eastern seaboard of the Greek mainland, in Attica, the Argolid and Thessaly, for whom obsidian became an important resource. Following Helms (1988), we might suggest that the fishing groups, by virtue of their knowledge of the Aegean islands and their access to resources not directly available on the mainland, acquired a particular status in relation to other mainland groups. Certain categories of material items associated with these groups, and specifically with their maritime travels, may then have acquired a symbolic significance in other communities, hence the prestige value of items such as marble vases and shell beads. Whilst Perlès (1992) is surely right, therefore, to point to the differentiation between the circulation of utilitarian lithics and symbolic prestige goods, the two may well have been intimately bound up with one another. Mainland communities' perceptions of the (as yet largely uncolonised) Cycladic Islands may well be central to an understanding of both processes. These networks of interaction, tied to long-distance fishing strategies, may have made a significant contribution to the economic, social and ideational climate within which the Aegean islands were colonised in the Late Neolithic and Early Bronze Age.

NETWORKS OF EXCHANGE IN THE NEOLITHIC OF THE CENTRAL MEDITERRANEAN ISLANDS

As in the Aegean, obsidian was a key resource throughout the Neolithic and Bronze Age of the Central Mediterranean and, as in the Aegean also, outcrops occur only on islands. Four central Mediterranean islands have been identified as significant sources of obsidian: Sardinia, Lipari, Palmarola and Pantelleria (Hallam *et al.* 1976). As in the Aegean again, however, the published literature on Neolithic exchange systems has tended to focus specifically on obsidian (Tykot 1992; Phillips 1992), perhaps to the exclusion of other materials. Malone (1985) attempts to put the obsidian exchange into a broader context, by looking at the circulation of a variety of materials in the area which includes Sicily, the Aeolian Islands and southern Italy. This area can be extended to include the Maltese Archipelago, and the islands of Pantelleria and Lampedusa (Stoddart *et al.* 1993). In this area, exchange seems to have involved the circulation of both stone and ceramic assemblages.

145

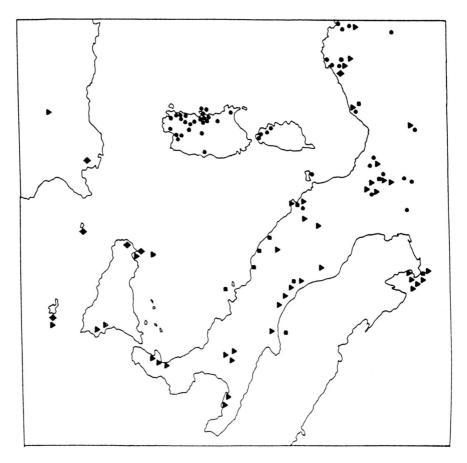

Figure 6.2 Map of central Mediterranean sites with obsidian tested by
chemical analysis

Source: Tykot 1992

Note: Triangles, Lipari obsidian; circles, Sardinian obsidian; squares, Palmarola
obsidian; diamonds, Pantelleria obsidian

In terms of quantity, obsidian was clearly the most important item of
exchange in the central Mediterranean area. Early research (cf. Hallam *et al.*
1976) suggested that, whilst Aeolian (Lipari) and Pontine (Palmarola) sources
supplied most of the obsidian found in southern Italy, Sardinian sources supplied most of the obsidian found in northern Italy and southern France.
The most recent work, however (Williams-Thorpe *et al.* 1979), suggests a
more complex pattern, with obsidian from Pantelleria, Lipari and Palmarola,

Table 6.1 Breakdown of Sardinian obsidian by source within distribution area (all figures shown as % of Sardinian assemblage within region)

Region	SA	SB	SC
Sardinia	37	6	58
Corsica	5	42	53
S France	94	0	6
N Italy	76	7	17

Source: Tykot 1992

as well as Sardinia, being found (albeit in small quantities) in the most remote corners of northern Italy. The distribution of obsidian from sources in the central Mediterranean is shown on Figure 6.2. It will be seen that, whilst assemblages from southern Italy are overwhelmingly dominated by Lipari obsidian, and assemblages from Sardinia and Corsica consist exclusively of obsidian from Monte Arci, assemblages from northern Italy include obsidian from all four sources.

As in the Aegean, the exchange of obsidian in the central Mediterranean seems to have begun at a relatively early stage. Recent survey in the Salerno Peninsula of southeast Italy identified obsidian (assumed to be from Lipari) in assemblages which, on typological grounds, would be assigned to the Late Upper Palaeolithic or Early Mesolithic (Milliken and Skeates 1989): it should be stressed, however, that these are surface finds, without a secure archaeological context. More securely, obsidian is found in levels 7 to 9 in the Uzzo Cave of Sicily, dated to the mid-sixth millennium cal. BC (Constantini *et al.* 1987). Characterisation studies (Francaviglia and Piperno 1987) have shown that both Lipari and Pantelleria obsidian are present in this assemblage. Obsidian is also present in the underlying levels (10 to 12) at Uzzo, though this is undated. The earliest date for Sardinian obsidian is from Basi (Corsica), where a single radiocarbon date calibrates within the mid-seventh millennium BC, though Phillips (1992) suggests that obsidian hydration dates (between 5548 and 4875 BC) for material from Su Carroppu de Sirri are likely to be more reliable (Michels *et al.* 1984). Studies of Early Neolithic (Stentinello) sites in southern Italy (Ammerman 1985; Ammerman and Andrefsky 1982) have suggested that some mainland sites may have had a specific role in producing obsidian blades from imported pre-cores.

Evidence for obsidian procurement has been most comprehensively studied in Sardinia (Tykot 1992). Hallam *et al.* (1976), on the basis of characterisation studies, identified three separate types of Sardinian obsidian (referred to as SA, SB and SC), all of which are likely to have their sources in the Monte Arci region (Puxeddu 1958). Several distinct working areas have been identified in the Monte Arci region (Mackey and Warren 1983; Tykot 1992).

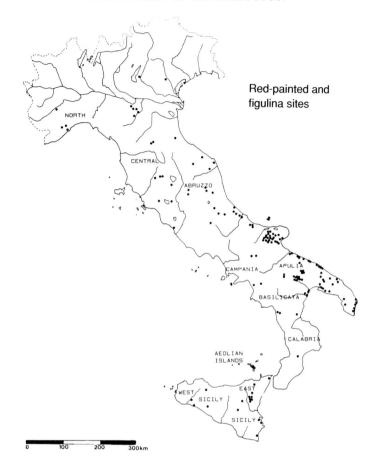

Figure 6.3   Distribution of Early Neolithic figulina and red-painted wares
Source: Malone 1985

Of these, Conca Cannas is clearly the source of SA obsidian, whilst obsidian of SC type is present in several localities in the Perdras Urias zone. Obsidian from a source at Cucru Is Abis is close, but not identical, to SB. All three types of obsidian seem to have been exploited from Early Neolithic times (all are present in Early Neolithic deposits at Su Carroppu) but, whilst SA and SC continued to be exploited into the Nuraghic period, there is no evidence for the exploitation of SB obsidian after the Late Neolithic (Tykot 1992). The distribution area of Sardinian obsidian includes Sardinia, Corsica, southern France and northern Italy. Table 6.1 shows the breakdown of Sardinian obsidian by source within these regions.

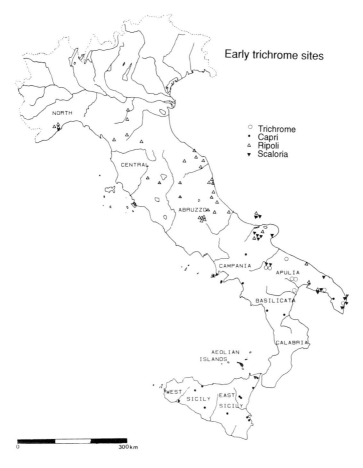

Early trichrome sites

○ Trichrome
• Capri
△ Ripoli
▼ Scaloria

Figure 6.4 Distribution of trichrome pottery styles

Source: Malone 1985

Clear distinctions are shown in Table 6.1 between the distribution patterns of obsidian from different sources in different regions. Whilst the SA source clearly supplied most of the Sardinian obsidian used in southern France and northern Italy, the SC source supplied most of the obsidian used within Sardinia itself, and also in Corsica. This suggests either that Sardinian and foreign communities both had direct access to the sources, or that procurement for long distance exchange was organised separately from local procurement. The position of the SB source is especially interesting: within Sardinia itself, the distribution of SB obsidian is extremely localised, occuring on a very limited range of sites, yet it is found in significant quantities in Corsica, and

Figure 6.5   Distribution of Serra d'Alto pottery
Source: Malone 1985

in smaller quantities in northern Italy. This could be taken as evidence supporting a direct access model, with procurement of SB obsidian being undertaken primarily by non-Sardinians (perhaps Corsicans).

Obsidian was not the only lithic resource to circulate within the central Mediterranean area. Stone axes, especially greenstone axes, are widely distributed in southern Italy, Sicily, the Aeolian Islands and the islands of the

150

Table 6.2 Quantification of Italian Neolithic finewares in ritual and domestic contexts

Style	% in ritual contexts	% in domestic contexts
Figulina	48	52
Red painted	51	49
Ripoli	54	46
Capri	48	52
Scaloria	60	40
Serra d'Alto	49	51
Diana	45	55

Source: Malone 1985

Maltese Archipelago (Evett 1973). Whilst it is possible that some of these axes have their origin in Alpine sources, Malone (1985) considers that most are probably derived from the crystalline rocks of the sila of Calabria. In the Maltese Archipelago (Stoddart *et al.* 1993), greenstone axes are a particular feature of assemblages of the Zebbug Phase (*c.* 4100–3800 cal. BC).

Pottery vessels appear to have circulated within clearly defined regional networks (Malone 1985), which included both mainland and island communities. As in the Aegean, ceramics are characterised by a high degree of 'stylistic investment' (Perlès 1992), suggesting that pottery may have been an important medium for the signalling of social and cultural identity. The depositional contexts of fineware bowls in the Italian Neolithic also suggests that pottery may have had an important role in ritual and funerary practice.

Early Neolithic (sixth and early fifth millennia cal. BC) assemblages are dominated by impressed wares: these include both coarsewares, which seem to have been produced and used on a purely domestic basis, and 'semi-fine' wares (Malone 1985) which may have circulated on a larger regional scale. Two types of fineware pottery are also present in the Early Neolithic – unpainted figulina and red-painted pottery – and these seem to have circulated within much larger networks of interaction and exchange. Although the distribution of these pottery types is clearly centered on Apulia, examples are also found in central and northern Italy, on Sicily and in the Aeolian Islands (Figure 6.3). Malone (1985) comments on the particularly high concentration of imported goods (especially southeast Italian finewares) on the island of Lipari.

From the mid-fifth millennium cal. BC onwards, several trichrome painted pottery styles can be distinguished in Italian assemblages, with clear regional distributions (Figure 6.4). The Scaloria style is concentrated in Apulia, but

Figure 6.6 Distribution of Diana pottery
Source: Malone 1985

is also found in the Tremiti Islands, alongside fineware vessels of Dalmatian types. The Capri style is concentrated in Sicily and southwest Italy, including the Aeolian Islands, and large quantities have been found on Lipari. The Ripoli style is concentrated in central Italy, but examples have also been found in northwestern Italy. Later trichrome pottery assemblages in southern Italy and Sicily are dominated by the Serra d'Alto style. Once again, the main concentration is in Apulia, but Serra d'Alto pottery is found throughout southern Italy, and also on the Aeolian and Maltese Islands (Figure 6.5). The Late Neolithic Diana style has its main concentration in Sicily, rather than Apulia (Figure 6.6), but again is found throughout southern Italy, with a significant concentration in the Aeolian Islands and examples also from Malta (Malone 1985).

The distribution of these pottery styles demonstrates that exchange networks between the Italian mainland, the Aeolian Islands and, to a lesser extent, Malta, involved the circulation of pottery vessels as well as lithics. It is, of course, possible that some of these pottery vessels served as containers for the circulation of perishable substances. Many of the fineware vessels circulating within these exchange networks were elaborately decorated, and their social significance is underlined by the contexts in which they are found. Table 6.2 shows the proportion of finewares in each style found in ritual (i.e. caves, burials, shrines) and domestic contexts.

The Aeolian Islands, particularly Lipari, seem to have played a key role in these exchange networks, both as a source of obsidian and as an importer of fineware pottery vessels and their contents (Bernabo Brea and Cavalier 1960, 1980; Malone 1985). Neolithic exchange networks in the central Mediterranean seem to have been fundamentally different from those in the Aegean. To begin with, the islands involved in these networks seem, in most cases, to have been fully colonised, which does not appear to have been the case in the Aegean. Island/mainland interaction, therefore, involved exchange in the strict sense rather than simply the procurement of raw materials by mainland groups. The social relations that were necessarily established between island and mainland groups seem to have taken a variety of different forms. The remarkable concentration of exotic items on Lipari has been noted both by Bernabo Brea and Cavalier (1960, 1980) and by Malone (1985). It is perhaps worth noting that the character of items imported to Lipari (elaborately decorated ceramic fineware, often deposited in ritual contexts) is in marked contrast to that of materials exported from Lipari (obsidian, used for utilitarian purposes, and found almost exclusively in domestic contexts). The pottery vessels are likely to have had a greater prestige value, and it is possible that social relations within Liparan communities were established and maintained through the control of access to these prestige goods, in a manner reminiscent of Melanesian Kula (Malone 1985). The Liparan elites, in turn, could use their control of economically important lithic resources to their advantage in allowing them to acquire mainland prestige items. Exchange between the Italian mainland (including Sicily) and the islands of Malta and Pantelleria may have had a similar social significance (Stoddart et al. 1993). Maltese communities may, for example, have controlled the flow of Pantelleria obsidian, whilst materials that had no particular prestige value on the mainland (greenstone axes, Sicilian ochre) may have acquired greater value in the context of a remote island group.

The situation in Sardinia is quite different, and seems not to have involved the circulation of prestige items (unless these were of a perishable nature). Sardinia has its own elaborately decorated Neolithic pottery styles (Bono Ighinu, Ozieri) and it is possible that this island's far greater size allowed the development of self-contained prestige exchange networks. The evidence for obsidian extraction and distribution (and particularly the differential distribution patterns of types SA, SB and SC obsidian) suggest that several modes of production and exchange coexisted with one another, and that procurement for local use was organised separately from procurement for long-distance exchange (Tykot 1992). It may well be that non-Sardinian communities had direct access to at least one of the sources in the Monte Arci region.

The islands of the central Mediterranean provide an interesting contrast with those of the Aegean, in demonstrating the variety of forms which island/mainland exchange systems may take. The obvious differences between these two case-studies may hide some fundamental similarities. Both involve the circulation of obsidian, found exclusively on the islands but much in demand in mainland communities. In both cases, obsidian itself seems to have been used for essentially utilitarian purposes, and had no particular social importance but, both in the Aegean and in southern Italy (though not apparently in Sardinia), obsidian circulated within networks of interaction that also included prestige items. In the case of the Aegean, it was particular materials from the coasts and islands which acquired prestige value in mainland communities, whilst in the Aeolian and Maltese Islands, it was mainland items that became socially significant in island communities. In both cases, however, the control of interaction *between* island and mainland communities was a significant factor in the articulation of social relationships *within* those communities, and this is a theme to which we will return.

INTERACTION AND EXCHANGE IN THE AEGEAN BRONZE AGE

Throughout most of the Neolithic period, the majority of the Aegean islands remained uncolonised, though they were clearly visited and their resources exploited by mainland communities. Crete is one of the few Aegean islands to show evidence for permanent settlement at an early stage in the Neolithic, and its anomalous position in this respect can probably be explained in relation to its size (see Chapter 3). The islands of Lemnos, Thassos (Leekley and Noyes 1975), Chios, Samos, Keos, Makronissos, Melos and Naxos (Hope-Simpson and Dickinson 1979) appear to have been colonised towards the end of the Neolithic, whilst a far greater number of islands (Hydra, Poros, Salamis, Spetsai, Lesbos, Nisyros, Delos, Despotiko, Donoussa, Heraklia, Ios, Keros, Kouphounissia, Rheneia, Schinoussa, Siphnos, Syros and Tinos) were colonised at the beginning of the Bronze Age, in the early third millennium cal. BC

(Cherry 1981). Inevitably, this dramatic activity in terms of island colonisation had a profound effect on the nature of inter-island and island/mainland exchange and interaction in the Aegean. The Early Bronze Age saw the development of complex exchange networks linking the Aegean islands to each other and to communities on the Greek and Anatolian mainlands. The most spectacular developments, however, took place in the Middle and Late Bronze Ages, when the Aegean area became the centre of a much larger international network, involving the exchange of objects and commodities from as far afield as Cyprus, the Near East and Egypt. The island of Crete seems to have played a particularly important role in these networks, leading some scholars to postulate the emergence of a Minoan 'thalassocracy', controlling the sea routes over the whole of the east Mediterranean region.

The Early Bronze Age (2700–2200 cal. BC)

The islands of the Aegean, particularly the Cyclades, remained an important source of raw materials throughout the Early Bronze Age. The circulation of Melian obsidian within the Aegean region continued, and reached its peak during this period (Torrence 1986). The Cyclades also have significant sources of copper (on Kythnos and Serifos) and silver (on Siphnos), and evidence for Early Bronze Age silver mining (Gale 1980; Gale and Stos-Gale 1981) has been identified at Aghios Sostis and Kapsalos (Siphnos). Interaction between the Cyclades and the Greek mainland also involved the circulation of marble figurines, from Paros or Naxos, and mainland pottery.

As in the Neolithic, obsidian seems to have served an exclusively utilitarian function, and had no special symbolic or social significance. Torrence (1986) has shown that the distribution pattern of Melian obsidian in the Bronze Age is significantly different from that in the Neolithic. Specifically, the Bronze Age pattern obeys Renfrew's (1977) 'law of monotonic decrement', which predicts that the quantity of a commodity should decrease with distance from its source, whilst the Neolithic pattern does not. Torrence interprets this as evidence for a decoupling of obsidian procurement and fishing strategies, and suggests that, for the first time, voyages were being made to Melos specifically for the purpose of obtaining obsidian. The social organisation of obsidian procurement and distribution in the Aegean Bronze Age has become a matter of some controversy. The original excavators of the site of Phylakopi (Bosanquest and MacKenzie 1904) argued that this site on Melos was involved in the large-scale commercial production of prismatic blades for export, a view which, in their opinion, was supported by the presence of a vast waste-heap of cores, chips and flakes, the so-called 'great obsidian deposit'. Torrence (1986) estimates the total volume of waste in this deposit at 7,289,700 grammes, corresponding to the production of 25 million blades. Whilst such a production would involve an estimated 94,918 labour days, Torrence points out that this would only represent the activities of a single worker over a period of 300 years. Given that the production of obsidian on the site continued for more than a millennium, the quantity of waste is not sufficient to support the hypothesis of full-time specialist obsidian working at Phylakopi. It is equally clear that the estimated 25 million blades produced at Phylakopi could not possibly have satisfied the demand for obsidian within the Aegean region. Torrence (1986) suggests that the 'great obsidian deposit' may simply reflect the manufacture of obsidian blades for local use, and points out that the production of blades from raw nodules or pre-formed cores seems to have taken place on virtually every Bronze Age site in the Aegean, casting doubt on the idea of completed blades being exported from Melos. Torrence (1986), like Renfrew (1984), favours a direct access model for obsidian procurement on Melos throughout the Bronze Age. Since Melos was, by this time, permanently settled, this would presuppose some form of agreement between the islanders and outsiders who wished to procure obsidian. It is possible that some form of payment was made to the islanders, in return for the right to procure obsidian, but it is equally possible that colonising groups maintained kinship bonds with their parent populations on the mainland over many generations, and that these mainland groups continued to exercise certain rights over the island's resources, as they had done before colonisation. Such a relationship may well have involved continued intermarriage between island communities and specific groups on the mainland, something which would, in any case, be necessary in the early stages of island colonisation, if the island population was to reproduce itself without incest.

The beginning of the Early Bronze Age (Early Cycladic I, *c.* 2700–2600 cal. BC) is marked by an increase in the production of marble objects, most notably figurines and vases. These were made on the islands of Naxos or Paros (where outcrops of marble occur, together with deposits of emery, used in the working of marble), and were exchanged within the Aegean area. More complex forms of figurine were developed in Early Cycladic II (2600–2300 cal. BC). These figurines (Figure 5.13) are found throughout the Cyclades, as well as on Euboia, and in the Attica region of the Greek mainland. Figurines are also widely distributed in central Crete, where local imitations were also produced. The context of deposition of these figurines suggests that they had some symbolic or ritual significance: the majority are from cist-grave burials, as at Chalandriani (Syros), whilst at Dhaskaleio Kavas (Keros) a large mass of broken figurines, stone vases and decorated pottery was found, in what appears to have been a votive deposition. The import of Cycladic marble objects to the Greek mainland can perhaps be seen as a continuation of Neolithic patterns of prestige exchange: a marble vase was found, for example, with a Neolithic child burial at the Francthi Cave (Perlès 1992). As in the Neolithic, significant quantities of Cycladic imports are concentrated on a small number of mainland sites, suggesting that certain communities had a privileged role in relation to island/mainland exchange. The site of Aghious Kosmas

(Mylonas 1959), on the Saronic Gulf near Athens, is perhaps one of these sites: a number of figurines were found in a cist-grave cemetery which also produced Cycladic-type pottery and stone vessels. Similarly, certain sites in Crete (notably Aghia Fotia and Archanes) have unusually high concentrations of Cycladic imports (Sakellerakis 1977; Davaras 1971).

Pottery also seems to have circulated within exchange networks of the Aegean Early Bronze Age, though the precise extent of this exchange has not been confirmed by petrographic or chemical analysis. In Early Cycladic I, pottery of Grotta-Pelos type forms a coherent assemblage, found on most of the Cycladic Islands: whilst there is no direct evidence for the exchange of this pottery between islands, the stylistic homogeneity of the assemblage testifies to the extent of inter-island contacts. In Early Cycladic II, pottery of Keros-Syros type is similarly homogeneous, but occurs alongside Urfinis wares (including 'sauceboats'), probably imported to the Cyclades from the Greek mainland. The evidence of the Dhokos Wreck (Koutsouflakis 1990) is particularly significant in understanding ceramic exchange in the Aegean Early Bronze Age: this vessel was carrying a cargo of around 1000 pottery vessels, along with millstones of andesite from Aegina. Metalwork was also increasingly significant, and demonstrates typological links with Anatolia, as well as with Greece.

The most remarkable aspect of Early Bronze Age exchange networks in the Aegean is the extent to which they represent a continuation and development of Neolithic practices. The direct access procurement of obsidian on Melos appears to have continued, as did the circulation of prestige items of marble from Paros and Naxos. As in the Neolithic, certain mainland communities seem to have had a privileged relationship with the Cycladic Islands. This is perhaps unsurprising since, in the early stages of island colonisation, one might expect island communities to remain bound by close ties of kinship to their parent populations on the mainland. The Early Bronze Age evidence corresponds very closely to the tripartite model of exchange systems developed by Perlès (1992) for the Aegean Neolithic (see above). The circulation of obsidian is likely to have had considerable economic importance to all communities, whilst marble objects seem to have had prestige status. As in the Neolithic, the exchange of pottery may have had a more localised social significance in establishing and symbolising inter-group alliances. Inter-island and island/mainland exchange in the Aegean Early Bronze Age seems to reflect a development of Neolithic practices and social institutions, in the changed circumstances brought about by the colonisation of the majority of the Aegean islands. Certain island commodities (e.g. marble) which had acquired prestige value during the Neolithic, retained that value in the Early Bronze Age, and access to marble objects was probably significant in the social reproduction of both island and mainland communities.

The island of Crete seems to have been involved in these exchange networks to a greater extent than it was during the Neolithic and, as on the Greek mainland, certain sites may have had a privileged involvement. The cemetery of Aghia Fotia, for example, produced unusually large quantities of Cycladic imports, whilst a tomb at Archanes produced thirteen marble figurines, three marble vases and nearly 1000 obsidian blades (Barber 1987). The greater involvement of Crete in inter-island exchange during the Early Bronze Age can perhaps be explained as a response to the colonisation of neighbouring islands (Karpathos and Kasos seem to have been colonised for the first time in the fourth millennium cal. BC, Kythera in the third millennium), which would inevitably have made Crete less isolated. A particularly close relationship may have developed between communities on Crete and those on Kythera (Coldstream and Huxley 1984). Whilst the earliest deposit of prehistoric material on the island (on Kastraki Hill) is of mainland character (with pottery of Early Helladic I–II type), a later settlement at Kastri has produced pottery with clear affinities in the Early Minoan II-II styles of western Crete. Identifiable Cretan imports are very few in number, and most of the ceramic assemblage seems to consist of local imitations of Cretan styles. The island has also produced a single Egyptian stone bowl, with an inscription dating it to the reign of Userkay (2487–2478 BC). Coldstream and Huxley (1984) consider the site of Kastri to represent a Minoan colony, but Rutter and Zerner (1984) point out that the earliest pottery is not associated with any architectural remains or burials, and suggest seasonal occupation by Cretan fishermen.

The final phase of the Early Bronze Age (Early Cycladic IIIA: 2300–2200 cal. BC) is marked by a significant decline in inter-island and island/mainland exchange. Melian obsidian continued to circulate within the Aegean region, but the production and exchange of marble figurines came to an abrupt end. New pottery styles appeared in the Cyclades, with clear Anatolian parallels, leading some authorities (cf. Barber 1987) to postulate an invasion from Asia Minor. The decline of exchange at this point in time effectively rules out any hypothesis which would see it as a 'prime mover' in the emergence of Cycladic towns, such as Phylakopi and Aghia Irini, since these appeared at a time when exchange was already in decline.

The Middle Bronze Age (2200–1700 cal. BC)

It is in the Middle Bronze Age that we begin to see evidence for the emergence of Crete as the major focus for cultural change and economic growth in the Aegean area. In the previous chapter we looked at the emergence of a 'palace civilisation' in the Middle Bronze Age of Crete: these developments will inevitably have created an increased demand for resources within Cretan society, and this demand is likely to have been met, at least in part, through exchange with other communities in the Aegean area and beyond. The extent of Cretan involvement in international trade in the Middle Bronze Age has perhaps been exaggerated in the past: one or two

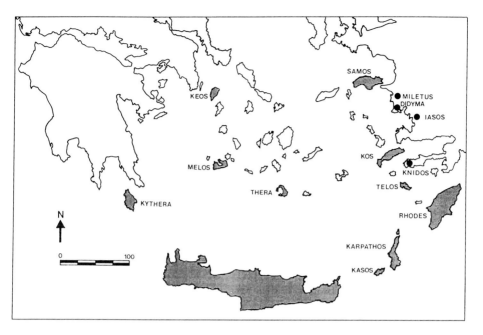

Figure 6.7 'Minoanised' sites in the Aegean

Source: Cherry 1981

imports, however spectacular, do not constitute evidence for regular contact or trade. Although the most dramatic explosion of exchange and trade in the Aegean area took place in the Late Bronze Age, the Middle Bronze Age evidence does suggest some significant changes to existing patterns of interaction within the Aegean.

For the first time, Cretan imports (mainly pottery) are found in significant quantities on the Greek mainland (as, for example, at Lerna), and on the Aegean coast of Anatolia (at Knidos, Iasos and Didyma). In some cases, as at Lerna, local imitations of Minoan pottery types are found alongside true Cretan imports. Particular concentrations of Minoan and 'Minoanising' pottery have been found on a small number of sites, suggesting highly directional trade (Dickinson 1994). Large quantities of both pottery types were found, for example, at Lerna, yet at the nearby settlement of Asine (Nordquist 1987) the proportion of imports is much lower, and 'Minoanising' pottery is entirely absent. Although Crete seems to have been a net exporter of pottery, some imported pottery is present at Knossos, both from the Cyclades (MacGillivray 1984) and, to a lesser extent, from the Greek mainland (Rutter and Zerner 1984). The extent of exchange between Crete, the Cyclades

159

and the Greek and Anatolian mainlands seems to have been relatively limited in the early stages of the Middle Bronze Age. In Middle Helladic I contexts, for example, Minoan pottery on the Greek mainland is confined to the eastern Peloponnese (Rutter and Zerner 1984), whilst in the broadly contemporary Middle Minoan IA horizons at Knossos, only a small quantity of Melian and central Cycladic vessels are present. The distribution of Minoan pottery on the Greek mainland was greatly extended in Middle Helladic II, to include the interior of the Argolid, along with coastal areas of eastern Attica and Thessaly (Rutter and Zerner *op. cit.*). On Crete itself, the quantity of Cycladic imports increased significantly in the later part of the Middle Bronze Age. MacGillivray (1984) records the presence of over forty Cycladic jars and amphorae in Middle Minoan III contexts at Knossos: since all of these are containers, probably for oil or wine, she suggests that Cycladic communities may have been paying tribute to the Knossos elite, in the form of agricultural or viticultural produce. She points out that, whereas there are over forty such jars from Knossos, and a few from Kommos, there are none from other sites in Crete, suggesting a specific relationship between Cycladic communities and the palace of Knossos. A similarly special role for Knossos is suggested by the production and distribution of Kamares ware (MacGillivray 1987). Kamares ware was produced on Crete, and was one of the most widely exported of Middle Minoan pottery types. MacGillivray (1987) has shown that its production was highly standardised and that, on Crete itself, different variants can be distinguished. Kamares ware from central Crete (the area around Knossos and Phaistos), for example, is distinctly different from that found in eastern Crete (around Mallia and Palaiokastro). All of the exported Kamares ware seems to have been of the central Cretan type. MacGillivray (1987) has also shown that the greatest concentrations of Kamares ware on Crete are in the palace stores, suggesting that the palaces had a key role in receiving, storing and redistributing this pottery, and whatever was contained in it. It seems likely, therefore, that by the end of the Middle Bronze Age, at least some Cycladic communities had become partially incorporated in the redistributive economy organised through the Knossos palace.

Although pottery is the most common recognisable item of exchange in the Aegean Middle Bronze Age, it may not have been the most important. The pots themselves may, in many cases, have been simply containers for the exchange of perishable commodities, such as wine, olive oil or honey. Metals are likely to have been of considerable importance in developing exchange relationships between Crete and the Greek mainland. Although the Middle Bronze Age of Crete is characterised by increasing wealth in terms of bronze and gold objects, and the island itself has no significant metal sources, and Stos-Gale and Gale (1984) have argued that Laurion, in southern Attica, and southern Laconia, in the eastern Peloponnese, is the most likely source for the copper, lead, silver and gold used in Crete. It may well be that the demand for metals was a key factor in encouraging the expansion of Cretan contacts within the Aegean region.

In comparison with other Aegean islands, Crete is relatively remote both from the mainland of Greece and from the coast of Anatolia. It is, however, connected to both via a string of 'stepping-stone islands'. It should not, therefore, be surprising that these islands played a significant role in contacts between Crete and the rest of the Aegean world. Niemeier (1984) has pointed out that the degree of Cretan influence in the Middle and Late Bronze Age is far greater on some islands than on others. In particular, he has identified a 'western string' (Thera, Melos and Keos) and an 'eastern string' (Kasos, Karpathos, Rhodes) of strongly 'Minoanised' islands (Figure 6.7). These are precisely the islands which one might expect to be used as 'stepping stones' between Crete, Attica in the west and southwest Anatolia in the east, and in both cases the islands concerned are separated by distances which could be covered in a day's sailing. A third axis of contact is provided by the island of Kythera, acting as a link with the important metal sources of Laconia in the Peloponnese.

Each of the islands of the western string has a major settlement: Akrotiri on Thera, Phylakopi on Melos, Aghia Irini on Keos. At the beginning of the Middle Bronze Age, these centres seem to have closer relationships with the Greek mainland than with Crete, perhaps, as Barber (1987) suggests, because the more backward mainland communities had a greater demand for Melian obsidian. The earliest unquestionably Cretan pottery in the Cyclades is from Phylakopi, and dates to Middle Minoan IA: the earliest Minoan pottery at Aghia Irini dates to Middle Minoan IB/IIA (Davis 1981). Imports from the Greek mainland (Grey Minyan ware) are also present at Phylakopi (Barber 1984), but at no point during the Middle Bronze Age are either Cretan or mainland imports numerous on the islands of the western string. On the contrary, ceramic assemblages are dominated by local fabrics (Cycladic white fabric and dark burnished wares) displaying both mainland and Cretan influences. Benzi (1984) has suggested that Minoan pottery was a rare luxury item on these sites.

The 'Minoanisation' of the eastern string seems to have started at around the same time. At Ialysos (Rhodes) pottery has been found with affinities in Middle Minoan 1B/II, and small quantities of Middle Minoan 1/II pottery have also been recovered at Palio Mitato and Lakos on Karpathos, and at Trapeza, Kefala and Tou Stamati Ta Laki on Kasos (Melas 1985). Benzi (1984) has pointed out that a high proportion of the assemblage from Ialysos, although of Minoan type, appears to have been locally produced, and he contrasts this with the assemblages from the western-string islands, which are dominated by local styles, with only a few Cretan imports. This contrast suggests to Benzi (*op. cit.*) that Ialysos, unlike Phylakopi or Aghia Irini, may have supported a significant population of Cretan immigrants.

A similar argument can be made in the case of Kythera, where the ceramic assemblage from the site of Kastri consists largely of locally made vessels, but in Minoan tradition (Coldstream and Huxley 1984). The site of Kastri is also interesting in that the town plan itself is very close to that of contemporary Cretan towns, such as Mochlos, Pseira and Gournia. A town drain was apparently provided as early as Middle Minoan 1B, something without parallel outside Crete. Minoan burial traditions are also evident at Kastri: for example, a pithos burial associated with pottery of Middle Minoan 1A. As in the case of Ialysos, it has been suggested (Coldstream and Huxley 1986) that Kastri supported a population of immigrant Minoans. Geographically, it is likely that Kythera's closest Cretan contacts were with the western region of Crete, around the palace of Khania.

It seems likely, therefore, that we are looking at three separate axes of contact between Crete and the outside world, probably dominated by different Cretan communities. The link between western Crete, Kythera and the metal sources of the eastern Peloponnese was perhaps controlled by the Khania palace (Rutter and Zerner 1984). The link between central Crete and southern Attica (including the metal source of Laurion) via the western-string islands was undoubtedly controlled from Knossos, whilst the link between eastern Crete and Anatolia, via the eastern-string islands was probably controlled by Mallia.

It would be a mistake to give the impression of Middle Bronze Age exchange networks in the Aegean being totally dependent upon Crete. The circulation of Melian obsidian continued throughout the Bronze Age, and there is no evidence to suggest that this ever came under any form of control from the Minoan palaces. Millstones of andesite from Aegina are also found increasingly on the Greek mainland (Dickinson 1994), whilst at the mainland site of Lerna, the ceramic assemblage includes Cycladic white painted ware, as well as Kamares ware from Crete. 'Duck vases' (Rutter 1985) are particularly interesting, since they are entirely absent from Crete, though they are found in the Cyclades, the eastern Greek mainland, the islands of Samos, Rhodes and Kalymnos and western Anatolia. Locally produced imitations of Cycladic duck vases are also found in Cyprus and Anatolia. The circulation of duck vases and Cycladic white painted ware as well as Melian obsidian may reflect the continuation of earlier patterns of exchange alongside the new interaction networks dominated by Crete.

Crete does seem to have had a monopoly over contacts between the Aegean world and areas to the east, though it is less clear whether these contacts occurred as a result of Cretan initiatives, or those of Near Eastern communities. Minoan imports (most notably Kamares ware) have been found on Near Eastern sites, but always in small quantities, though interestingly, local copies of Kamares ware have been found on the Egyptian sites of Kahun and Haragch (Dickinson 1994). Near-Eastern imports to Crete include Egyptian scarabs and stone vessels, Syrian daggers and a jar of Early Cypriot

III type from Knossos (MacGillivray 1984), but again, these occur in very small quantities, certainly not in themselves sufficient to suggest regular trade. Texts from Mari, dating to the eighteenth century BC, refer to goods from 'Kaptara' (usually assumed to be Crete), apparently gifts to the local King, from the King of Ugarit, where a group of Kaptaran merchants was based (Heltzer 1988, 1989). Assuming that the identification of Kaptara with Crete is correct, a whole list of Minoan products is given, including weapons, textiles, pottery and sandals, but interestingly, no gifts are recorded as having been sent by the King of Ugarit to Kaptaran rulers, suggesting that Crete did not rank particularly highly in the international politics of the period. There is some evidence to suggest that the limited contacts which did take place between Crete and the Near East may have been controlled by the Knossos palace, in that all of the Kamares ware found in the Near East and in Egypt is of the central Cretan type (MacGillivray 1987).

The Middle Bronze Age marks a significant change in the organisation of exchange networks in the Aegean. Evidence for contact with the Near East, though spectacular in terms of its contrast with earlier periods, is not necessarily the most important aspect of this change. The small quantities of material involved could simply reflect occasional visits to Crete by Near Eastern trading ships, though the Mari texts do hint at the activities of a small number of Minoan merchants abroad. Some aspects of exchange in the Aegean Middle Bronze Age can be seen as evidence for a continuation of Early Bronze Age and even Neolithic traditions: the circulation of Melian obsidian and Aeginitan andesite, for example, seems to have continued as before, unaffected by developments in Crete. Although a distinctively new pottery form, 'duck vases' (Rutter 1985) seem to have circulated within networks of interaction, the origins of which can be traced back much earlier, and from which Crete was excluded. The appearance of copies of these vases in Anatolia and Cyprus probably has more to do with the activities of Near Eastern and Anatolian traders than with those of Cycladic islanders. Arguably the most important development of the Aegean Middle Bronze Age in terms of exchange and trade was the incorporation of most of the Aegean islands, together with parts of the Greek and Anatolian mainlands, in a regional economic system which had its core in Crete. As in all 'core–periphery' situations, the net flow of resources within this system was centripetal, from periphery to Cretan core. Metals from the Greek mainland were perhaps the most important resources in this system (Wiener 1990), and the development of the Minoan 'palace civilisations' in Crete would certainly have created considerable demands for copper and gold in particular. Agricultural surpluses may also have been involved (as suggested by the Cycladic storage vessels found at Knossos) and the Aegean islands may have become increasingly drawn into the tribute-based redistributive economies operated by the Minoan palaces. The economic and political success of emerging Cretan polities depended, to a large extent, upon their ability to ensure this centripetal flow of resources from a continually expanding hinterland. The establishment of 'Minoanised' settlements on Kythera, and on the islands of the western and eastern strings can be seen as a response to this need. In some cases (as, perhaps on Kythera and Rhodes) this may have involved the deliberate colonisation of an island, whilst in other cases (such as Melos and Keos) it is more likely to have involved the establishment of alliances with native populations, with promises of luxury goods in return for raw materials. In either case, the geography of the Aegean is likely to have made this process easier. Islands, being bounded and relatively small, are easy to control and defend. The islands of the southern Aegean form a necklace linking Greece and Anatolia, with Crete at its centre. This configuration makes it relatively easy to control the sea routes linking Crete with the continents of Europe and western Asia: the short distances between islands minimise the dangers of navigation, both from natural risks and from piracy (assuming that all the islands in a chain are effectively controlled). By maintaining influence (either through coercion or, more likely, through networks of alliances) over a very small number of people in a few small islands, it would thus be possible for the Minoan elites to establish effective control over the trade routes linking Crete to the eastern and western shores of the Aegean Sea. The Minoan 'thalassocracy', therefore, can be seen as a specific sociopolitical manipulation of the geographical realities of the south Aegean archipelago. Whilst it was in no sense a 'prime mover' in the emergence of the Minoan palace civilisations (the most dramatic expansion of trade and exchange took place in the Late Bronze Age, long after the appearance of the first palaces), this strategic control of a few islands may have played a significant part in their expansion.

The Late Bronze Age (1700–1200 cal. BC)

The Late Bronze Age is marked by a dramatic expansion of Minoan influence throughout the Aegean, and by increasing evidence for contacts between the Aegean (particularly Crete) and the civilisations of Egypt and the Near East. The period is also marked by the emergence of wealthy centres on the Greek mainland, which became increasingly involved in trade and exchange with Crete.

As in the Middle Bronze Age, the spread of Minoan influence is marked by pottery styles, and Minoan fine wares and domestic pottery (both Cretan imports and local copies) did become increasingly common throughout the Aegean during the Late Bronze Age. The 'Minoanisation' of the Aegean, however, involved far more than simply the spread of pottery styles. The Cretan weight system and Linear A script were increasingly used elsewhere in the Aegean. A Minoan loom type was adopted, for the first time, in the Cyclades and, on some islands, Minoan architectural styles were introduced (Dickinson 1994).

The effects of Minoanisation were by no means even across the Aegean region. On the Greek mainland, Cretan imports are found in significant quantities, mostly in burial contexts. Much of the material from the shaft graves at Mycenae is Cretan in origin, whilst the grave goods from a wealthy tomb at Vapheio include a set of balances and a Minoan lead weight (Dickinson 1994). The burials themselves, however, are distinctively Helladic rather than Minoan in character, as are the associated settlements. In the Cyclades, Minoan imports are largely confined to Kythera, and to the islands of the western and eastern 'strings': elsewhere, apart from a handful of Cretan imports from Naxos and Delos, the assemblages are essentially Cycladic. The major settlement of Paroikia (Paros), for example, has no Cretan features and no Minoan imports (Davis 1981). Pottery continued to be made in established Cycladic traditions, and is quite widely distributed: the assemblage from the second palace at Knossos includes Melian, Theran and possibly Naxian pottery (MacGillivray 1984). The circulation of Melian obsidian declined slowly as metal objects became more widely available, but it continued throughout most of the Late Bronze Age. Contacts between Aegina and the Greek mainland also continued: pottery from Aegina has been found on a range of mainland sites, whilst Helladic Minyan ware is found on Aegina and other Cycladic islands (Dickinson 1994).

The most spectacular developments are on the islands of the western string – Thera, Melos and Keos, and the contrast between these and other Cycladic islands is truly striking, leading some authorities to postulate the existence of Minoan colonies (Branigan 1981, 1984; Wiener 1984). The close relationship between these islands and Late Minoan Crete is demonstrated by material culture, architecture and religion.

The towns of Akrotiri, Phylakopi and Aghia Irini have produced large quantities of Cretan imports. As well as pottery, Minoan imports include lead weights and tablets with Linear A inscriptions (Barber 1987). Material from the western-string islands is also found both in Crete and on the Greek mainland: large quantities of Melian pottery are present at Knossos, whilst Theran 'bird jugs' are present in Grave Circle B at Mycenae (Barber 1987). Both Theran and Melian pottery have been identified on sites in Attica and the eastern Peloponnese (Davis 1981), and mainland pottery is present in quite large quantities at Aghia Irini (Barber 1987). In addition to exchange with Crete and the Greek mainland, significant exchanges seem to have taken place between the three main islands of the western string: Melian pottery is present both at Akrotiri and at Aghia Irini, whilst Theran pottery is also present at Aghia Irini and Phylakopi (Davis 1981). Melian pottery is present on the islands of Naxos, Siphnos, Delos and Tinos, demonstrating that exchange took place between the western string and other Cycladic islands (Schofield 1982). Occasional items from further afield seem to have reached the islands of the western string, presumably via Crete: Cypriot sherds are present on both Thera and Melos, whilst the assemblage from Akrotiri includes Syro-Palestinian stone vessels and a plain Syrian amphora (Barber 1987). Despite the presence of imports, much of the locally made pottery in the western-string islands is distinctively Cycladic in character, arguing against the suggestion that the population consisted largely or predominantly of immigrant Minoans: at Phylakopi, for example, almost all of the pottery (apart from imports) is of local type (Barber 1984), though the assemblage from Akrotiri is very different, with local copies of Minoan forms predominating (Doumas 1980).

Cretan architectural features on the islands of the western string include the use of ashlar masonry, Cretan room units such as lustral basins, poly-thyra and pillar crypts, and the existence of figured fresco scenes. These features are most clearly seen on the site of Akrotiri (Marinatos 1984a and b). The large independent 'Xeste' buildings (Figure 5.19) have parallels in the Cretan 'villas', and are quite unlike anything found elsewhere in the Cyclades. Xeste 3 has a polythyron and a lustral basin, both typically Minoan features. Several of the buildings at Akrotiri have frescoes, the iconography of which links them clearly to Minoan Crete and, more specifically, to Knossos (see Chapter 5). At Phylakopi, two pillar crypts have been identified, one of them with a fresco depicting a robing ritual (Dickinson 1994). The architecture on the western string islands, however, is by no means identical to that of Crete. Houses of unmortared drystone blocks at Aghia Irini are certainly not built in the Minoan style, and even at Akrotiri the buildings have some distinctive local features, such as the presence of vertical wooden supports in interior walls (Branigan 1984).

Evidence for Cretan religious traditions in the islands of the western string is most clearly seen at Akrotiri, where many of the frescoes have a clearly religious character and obvious parallels at Knossos (Marinatos 1984 a and b; see also Chapter 5). There are other points of comparison between cult activities at Akrotiri and those of the Minoan palaces: these include the association of ritual activity and food preparation, and the presence of 'cult gear', such as rhyta, offering tables and horns of consecration. The 'temple' structure at Aghia Irini is in no sense a Minoan building, but it has produced fragments of at least fifty-five female figurines of typical Minoan style. The nearby site of Troullos Hill has certain features in common with the Minoan peak sanctuaries, and finds include two offering tables and a bronze figurine (Davis 1984). Wiener (1984) has drawn attention to the 'super-abundance' of conical cups on the sites of Akrotiri, Phylakopi and Aghia Irini: a single house at Aghia Irini (House A) produced over 8000 of these vessels, including 820 from a single deposit beneath Room 1B. Contextual evidence, both in Crete and in the western-string islands, suggests that these cups served some form of ritual function: they occur in large quantities at Cretan peak sanctuaries and in a votive deposition in the Diktaian Cave, whilst at Aghia Irini, conical cups are built into the breasts of terracotta figurines. The presence of such large numbers of conical cups on the western-string islands

Figure 6.8 Figures representing Cretans bearing tribute, from the Tomb of Senmut at Thebes

Source: Sakellerakis and Sakellerakis 1984

suggests to Wiener (*op. cit.*) that Cretan-style rituals were being enacted on these islands.

The archaeological evidence from the western string islands suggests a continuation and intensification of the patterns of interaction established in the Middle Bronze Age. Cretan interest in the supply of metal from Attica is likely to have been a key factor in these developments, and Wiener (1984) has drawn attention to the evidence for metal working at Aghia Irini: large numbers of crucibles were found in Level VII (Late Minoan IB), including seventeen from House A. Other commodities which may have circulated within this network (Davis 1981) include wool (study of the faunal assemblage from Phylakopi suggests the intensive raising of sheep for wool) and saffron (the gathering of which is shown on a fresco from Akrotiri).

The islands of the western string are not unique in showing evidence for intensive exchange relationships with Crete. The Late Bronze Age assemblage from Kastri (Kythera) is dominated by Cretan imports and local imitations. Despite this island's proximity to the Greek mainland, Helladic imports did not appear in any significant quantity before Late Minoan 1B, and even then were vastly outnumbered by Minoan imports (Coldstream and Huxley 1984). Late Bronze Age burial practices on Kythera also follow Cretan patterns, with natural caves adapted to create collective tombs. The evidence from the islands of the eastern string is less well understood, though this is probably due to a lack of research rather than a lack of evidence. Large quantities of Minoan conical cups have been found both on Tilos and Kos (Niemeier 1984).

Whilst it seems clear that patterns of interaction in the Aegean Late Bronze Age were directional, and served to ensure a centripetal flow of resources towards Crete, it is less clear by what means Cretan communities exercised such a strong influence over Kythera and the islands of the western and eastern strings. The extent of Minoanisation on these islands has led some authorities to identify them as Cretan colonies. Branigan (1981) has distinguished three types of colonialism. Governed colonies (such as India under the British Empire) are settlements of largely indigenous people, controlled and governed by a foreign power: there is no evidence for colonies of this kind in the prehistoric Aegean. Settlement colonies (such as the earliest European colonies in North America) are settlements of immigrants, which may either be independent or controlled by the home power. Branigan suggests that Kastri (Kythera) may have been a settlement colony, since most of the material culture (even locally produced pottery) is Minoan in character. Community colonies (such as the Chinese community in Bankok) are minority immigrant populations within communities ruled by indigenous groups. Branigan (1981) argues that the Minoanised settlements of Akrotiri, Phylakopi and Aghia Irini are most likely to have been community colonies. It is certainly true (Marinatos 1984 a and b) that the material culture, architecture and religious practices of these communities show distinctive local features, and this makes it unlikely that immigrant Minoans were in a majority. It could legitimately be asked, however, whether it is necessary to invoke colonialism at all as an explanation for the appearance of Minoan features on the islands of the western and eastern strings. Wiener (1984) argues that Minoan architecture, art and material culture may have been copied by local elites within the Aegean islands, as a means of enhancing their own status, a phenomenon which he describes as the 'Versailles effect'. It is a common feature of core/periphery situations that contact with the core leads to increasing levels of hierarchy in the communities of the periphery, and that social relations within these peripheral communities become articulated through the control of relations with the core area. Local elites on Thera, Melos and Keos may simply have turned Cretan strategic interests in these islands to their own socioeconomic advantage, establishing alliances with Crete in return for access to prestige goods, and perhaps also the services of Cretan architects, artists and ritual specialists. This effect is likely to have been most clearly marked in areas close to the core: thus the strongest evidence for Minoanisation is on Thera, whilst in mainland Greece, on the very edge of the peripheral area, Cretan prestige goods were socially significant, but Minoan architecture and religion seem not to have been adopted.

Interaction between Crete, the Near East and Egypt increased markedly during the Late Bronze Age. Imports to the Aegean include gold, ivory, ostrich eggs, copper and tin oxhide ingots, Egyptian alabaster and lapis lazuli from central Asia. Near Eastern pottery is also present in Crete, including Cypriot and Canaanite wares. Analysis of the oxhide ingots shows them to be non-Aegean in origin (Stos-Gale and Gale 1984): some are Cypriot, but a higher proportion are from an unknown source, possibly in Anatolia. Throughout most of the Late Bronze Age, the only Aegean exports to reach the Near East and Egypt were Cretan, though Mycenean pottery and swords of Late Helladic IIA/B have been found in Cyprus and Egypt. Cretan influence in the Near East is most clearly seen in Cyprus, where the Cypro-Minoan script seems to derive from Linear A. In addition to the archaeo-logical evidence for interaction between Crete and the Near East, Egyptian texts and tomb paintings give clear indications of the overseas activities of the Minoans (Sakellerakis and Sakellerakis 1984). It is generally considered that references in the Egyptian texts to the 'Land of Keftiu' relate to Crete. Official records for the thirty-fourth year of Thothmes III (1470 BC) refer to the 'ships of Keftiu' transporting Palestinian timber to Egypt. Relationships between Crete and Egypt seem to have been particularly close during the reign of Thothmes III (1504–1450 BC). Figures which probably repre-sent Cretans are shown in frescoes from the tombs of User-Amon (c. 1476 BC), Rechmere (c. 1470 BC), Menheperesenb (1504–1450 BC), Senmut (c. 1493 BC) and Puemre (1490–1480 BC). These figures carry gifts (Figure 6.8), some of which are clearly of Aegean origin, such as Minoan amphorae, rhyta and bull figurines, but others of which were not produced in the Aegean, including elephant tusks, lapis lazuli, silver and gold. It may be that Cretan maritime expertise allowed them to play an important (and lucrative) role as middle-men between the civilisations of Egypt and the Near East, something which is clearly hinted at in the official records of Thothmes III.

The collapse of Minoan 'thalassocracy'

The Cretan domination of exchange both within and beyond the Aegean came to an end in the mid-fifteenth century BC, with the close of the Second Palace period (see Chapter 5). Wealthy centres on the Greek main-land became increasingly powerful during the Late Bronze Age, and they seem to have responded to the Cretan control of trade and exchange within the Aegean by opening up a new axis of contacts with the central Mediter-ranean. Mycenean pottery (including domestic wares) of Late Helladic I has been found in the Aeolian Islands (Re 1986), whilst pottery of Late Helladic II has been found on a range of sites in southern Italy and (in one case) Albania (Dickinson 1986; Knapp 1990). Mycenean pottery begins to be found on a large scale throughout the Aegean in Late Helladic IIIA1. Later Mycenean pottery (Late Helladic IIIA2/B1) occurs in far greater quantities, however, and is found as far afield as Cyprus, the Syro-Palestinian coast, the central Mediterranean including Sardinia, the Anatolian coast and central Macedonia (Dickinson 1994). This period did not see a decline in international trade, the scope of which is demonstrated by the shipwrecks at Ulu Burun and Cape Gelidonya: the Ulu Burun ship was carrying a mixed cargo, from Egyptian, Near-Eastern, Cypriot and European sources (Bass 1986; Knapp 1990), whilst the Cape Gelidonya cargo included an Italian sword and Aegean stirrup jars, as well as a quantity of Cypriot material (Bass 1967, 1991). This trade, however, was no longer dominated by Crete, and the Aegean material found on these two shipwreck sites is Mycenean rather than Minoan. The decline in Minoan involvement begins in Late Minoan II/IIIA1, though it did experience a brief revival in Late Minoan IIIB, when Cretan influences became significant again in the Cyclades and in Cyprus (Hankey 1979).

The decline in Minoan influence abroad must be understood in the context of a series of major upheavals which took place in the Aegean in the fifteenth century BC, and which also involved the destruction, not only of most of the Minoan palaces, but also of most of the Minoanised sites on the Aegean islands. This may have involved both internal conflicts and Mycenean military expansion (Niemeier 1984), but it is clear that the ultimate beneficiaries were the Myceneans (see Chapter 5). The rebuilding of Phylakopi in Late Cycladic III included the construction of a classic Helladic megaron building, and a shrine which has produced a terracotta figurine of clearly Mycenean form, the so called 'Lady of Phylakopi' (Renfrew 1985). As the role of Crete in international contacts declined, so did the role of Kythera, and the islands of the western and eastern strings: this is unsurprising, since the wealth of these islands depended entirely on their geographical position as 'stepping stones' between Crete and the continents of Europe and western Asia.

The Myceneans did take a particular interest in certain Aegean islands, though for entirely different reasons. The island of Delos was later to become one of the most important cult centres of the Aegean during the Iron Age and Archaic periods, associated in particular with the cult of Apollo. Beneath the Temple of Artemis at Delos was found a votive deposition dating to Late Cycladic III (Barber 1987): this deposition included gold jewellery, miniature birds, bees and animals, Mycenean animal figurines and Mycenean or Cypro-Mycenean ivory plaques. The deposition suggests that the development of Delos as a cult centre has its origins in the Late Bronze Age: this is not entirely surprising, since we already know from Linear B inscriptions that the Olympian pantheon of Gods and Goddesses has roots in Mycenean religious belief. A consideration of the importance of Delos as a cult centre brings us back to Helms' (1988) concept of 'cosmological zoning'. In all societies, elements of the natural landscape are invested with symbolic and cultural meaning. Small islands, being bounded and, to some extent, inaccessible, lend themselves particularly to this form of categorisation, and the island of Delos seems to have been invested with cosmological significance in the Late Bronze Age. The small island of Samothrace acquired similar cosmological significance, though the earliest relevant finds date to the seventh century BC (Burkert 1993). Although a unique and, in some respects, extreme case Delos may perhaps give us some insights into the way in which particular islands may have been perceived at certain moments in time. This in turn may help us to understand the cultural significance which inter-island and island/mainland interaction may have had in prehistory.

It is also interesting to note that several of the Aegean islands acquired renewed significance in the trade routes between the eastern and western Mediterranean established by the Phoenicians and Cypriots during the Iron Age. Euboia became an important trading centre during the Greek 'Dark Ages', and Phoenicean imports dating to the tenth century BC have been found on the site of Lefkandi (Negbi 1992). Cypro-geometric pottery of the eleventh century has been found on Rhodes (Coldstream 1988). On Crete, large quantities of Cypriot and Phoenician vessels have been found in tombs in a cemetery of the tenth century BC at Knossos, alongside protogeometric vessels from Attica, Euboia and the Cyclades. The activities of Near Eastern traders in the Aegean islands are also reflected in the architecture of Iron Age temples and shrines (Negbi 1992) at Kommos (Crete) and Phylakopi (Melos). It seems that Near Eastern traders were using the Aegean islands as stepping stones, much as the Minoans had in the Middle and Late Bronze Age, and as offshore centres for commercial activities. Cypriot and Phoenician pottery is found on a very small number of Aegean islands (Rhodes, Crete, Melos, Euboia), and these are precisely the islands which are most likely to have been significant to trade routes linking the Aegean area with Anatolia, Cyprus and the Near East.

THE MYCENEAN AND PHOENICIAN PRESENCE IN THE ISLANDS OF THE CENTRAL MEDITERRANEAN

Mycenean communities responded to the Cretan monopoly over contacts between the Aegean and the eastern Mediterranean by establishing new trade links with the communities of the central Mediterranean, most particularly southern Italy, Sicily and Sardinia. Offshore islands, such as Lipari, and even tiny islets such as Vivara, in the Gulf of Naples, became important focal points for contact between the indigenous populations of Italy and visiting Mycenean traders (Bietti-Sestieri 1985). At a later stage, the islands of the central Mediterranean became important focal points for similar contacts between Italian communities and Phoenician traders (Negbi 1992). The smaller islands probably functioned as offshore trading centres: neutral territory on which commercial transactions could take place, unencumbered by the rules and conventions which governed such relationships in Italian mainland communities. This would work to the benefit of both foreign traders

171

and local elites, rather in the way that offshore finance centres operate in the modern capitalist world. Larger islands, such as Sardinia, are likely to have been chosen because of their strategic position, as 'stepping stones' between continental areas (e.g. between Italy, southern France and Iberia).

Mycenean contact with Italy is concentrated in four areas: the Ionian coast of Apulia, Eastern Sicily, the Aeolian Islands (Bietti-Sestieri 1985; French 1985) and Sardinia (Ferrarese-Ceruti 1979). Evidence for this contact includes significant quantities of imported Mycenean pottery, though local copies are also present in some cases, as on Sardinia (Ferrarese-Ceruti *et al.* 1987; Jones and Day 1987). Mycenean influences in architecture have been noted only on Sicily: rock-cut tombs of Mycenean type have been identified at Thapsos, Plemmyriai, Molinello and Cozo Pantano (Orsi 1892, 1896), and a large palatial complex at Thapsos has features suggestive of Mycenean influence (Voza 1972, 1973).

The great concentration of imported Mycenean material on the tiny islet of Vivara in the Gulf of Naples is particularly interesting, since Mycenean sherds are rare in adjacent mainland areas (Bietti-Sestieri 1985). This suggests that the islet may have played a special role in contact between Mycenean and Italian communities. Large quantities of Mycenean imports are also found on the Aeolian Islands, especially Lipari and Filicudi (Taylour 1958; Vagnetti 1983), and Bietti-Sestieri (*op. cit.*) suggests that the Myceneans may have made use of existing networks of contacts between the Aeolian Islands, Sicily and the Tyrrhenian coast. The Myceneans may have used the Aeolian Islands in much the same way as the Minoans used the islands of the western and eastern Aegean 'strings', establishing close alliances with small island groups as a means of gaining access to wider networks of contacts. The architectural evidence and material culture from the Aeolian Islands suggests native settlements rather than Mycenean colonies, though the evidence from Thapsos suggests that some Myceneans may have been resident in Sicily (Coles and Harding 1979).

The pattern of contacts between Sardinia and the Mycenean world is rather different. Mycenean pottery (Late Helladic IIIB) has been found on a number of sites, most notably Nuraghe Antigori (Giardino 1992), but petrological analysis shows that much of this was produced locally (Ferrarese-Ceruti *et al.* 1987; Jones and Day 1987), whereas most of the Mycenean material found elsewhere in the central Mediterranean appears to have been imported. Mycenean interest in Sardinia may have been prompted by the availability of metals: copper, silver, lead and iron are naturally present on the island (Marcello *et al.* 1978), and the copper and silver sources were exploited from Late Neolithic times (LoSchiavo 1989). Sardinia became a major centre for the production of metal objects during the Nuraghic period, as shown by the slag deposits found at sites such as Nieddue (Zwicher *et al.* 1980), Genna Maria Villanovaforru (Atzeni *et al.* 1987) and Sa Sedda 'e Sos Carros (LoSchiavo 1988). Mine shafts have been identified near the mouth of the Rio Sarraxinus, with pestles, mortars and heaps of burnt stone (Rellini 1923), and moulds for metal objects have been found on a number of sites (LoSchiavo 1982). Given the importance of Sardinia as a source of metal ores, it is perhaps surprising that among the most significant imports to the island from the eastern Mediterranean are copper ingots of the 'oxhide' form. Oxhide ingots of the Late Bronze Age have been found in Cyprus, Greece, Crete, Sardinia, Sicily and Lipari (Muhly *et al.* 1988). Chemical analysis suggests that they were made of Cypriot copper (Gale and Stos-Gale 1986, Stos-Gale and Gale 1992), though the only mould as yet discovered is from the site of Ras Ibn Hani in Syria (Lagarce *et al.* 1983). Sardinia has the largest concentration of oxhide ingots found on land: a total of fifty are recorded from Sardinia, compared with thirty-seven from Crete, twenty-two from mainland Greece and twenty from Cyprus (Stos-Gale and Gale 1992). Bass (1986), however, points out that these figures are insignificant compared to the 200 found as part of a single cargo on the shipwreck of Ulu Burun. Lead isotope analysis suggests that, whereas the oxhide ingots are probably made of Cypriot copper, those Nuraghic bronzes which have been analysed show an entirely different composition, compatible with the use of local ores (Stos-Gale and Gale 1992). Oxhide ingots from Sardinia have commonly been found alongside 'bun-shaped' copper ingots, as at Villanovaforru, Villagrande, Strisaili and Ittireddu and, where these have been analysed, they are of local rather than Cypriot copper (Stos-Gale and Gale *op. cit.*) Interestingly, the association of oxhide and bun-shaped ingots is also found on the shipwrecks of Cape Gelidonya and Ulu Burun. The existence of a relatively small number of east Mediterranean copper ingots on Sardinia (around 25 per cent of one cargo, judging from the evidence of the Ulu Burun wreck) may disguise the fact that the net flow of metals was from Sardinia to the east Mediterranean and not *vice versa*. Cypriot copper imported to Sardinia seems not, on the evidence of the available analyses, to have been used in the production of local bronzes. On the contrary, imported ingots were kept intact, possibly having some significance, either as prestige items in their own right, or as a form of currency. The presence of oxhide ingots in hoards, for example beneath the floor of Nuraghe Sa Antioco di Bisarcio (LoSchiavo 1984), lends some weight to the suggestion that the oxhide ingots may have had a value greater than that of the raw material. This returns us yet again to Helms' (1988) suggestion that value is often added to objects by virtue of a distant provenance. If items imported from the eastern Mediterranean acquired prestige value in Sardinian communities, it seems that Sardinian objects acquired a similar value in the communities of the Italian mainland. Gras (1985) draws attention to the existence of nuraghic figurines, daggers and pins in graves of the Villanovan and orientalising phases along the Tyrrhenian coast of Etruria. Giardino (1992) argues that these should be seen as votive offerings, a suggestion supported by the depositional contexts of figurines in Sardinia itself, where they frequently

173

occur in hoards, sometimes associated with well sanctuaries (Lilliu 1975), as at Santa Vittoria di Serri.

Following the decline of Mycenean power in the twelfth century BC, the Phoenicians became the dominant maritime nation in the eastern Mediterranean (Negbi 1992), establishing a colony at Kition in Cyprus in the mid-ninth century BC (Karageorghis 1976). Euboia and Crete became important 'stepping stones' on Phoenician trade routes linking the eastern and central Mediterranean. Within the central Mediterranean, the main areas of Phoenician contact were in Sicily (the site of Motya) and Sardinia (Sulcis and Tharros), and these islands seem to have maintained their pivotal position in Mediterranean trade throughout much of the Iron Age.

PATTERNS OF INTERACTION IN MEDITERRANEAN ISLAND PREHISTORY

Helms (1988) has shown how geographical space is commonly accorded cultural significance, and how this can be manipulated by elites through control of interaction with distant areas. Islands, being clearly defined and accessible only by boat, offer particular opportunities for these forms of social practice. Power and prestige may depend to a large extent upon participation in inter-island and island/mainland exchange, mediated through elite control of overseas contacts, access to boats and navigational knowledge, as in the case of Melanesian Kula (Malinowski 1922; Munn 1986; see also Chapter 2).

Island/mainland exchange may be a significant factor in defining social relationships within both island and mainland societies. In the Aegean Neolithic, for example, the control of interaction with the islands may have been an important factor in the mediation of social relations within and between mainland communities. The concentration of island imports (obsidian, stone vases, etc) on a few mainland sites, such as Francthi, suggests that contact with the islands was controlled by a limited number of mainland groups, which probably specialised both in large-scale fishing and in obsidian procurement (Perlès 1992). The islands, most of which were not colonised until a very late stage in the Neolithic, may have had particular significance in the 'sacred geography' of Aegean communities, in which case certain objects and materials from the islands may have acquired symbolic or prestige value, as seems to have been the case with stone vases and figurines. Obsidian, by contrast, seems to have had no particular symbolic significance, though the control of circulation of this key resource would probably have enhanced the status of those groups which engaged in its procurement. These patterns of island/mainland interaction continued into the Early Bronze Age, with the circulation of marble figurines from Naxos or Paros: once again, a small number of mainland sites (such as Aghios Kosmas on the Saronic Gulf) seem to have had a privileged access to prestige goods from the islands.

The importance of particular islands in the cosmology of Aegean communities is a theme which runs through the prehistory and antiquity of the region, as witness the importance of sanctuaries on Delos and Samothrace in the religion and mythology of the Archaic and Classical periods.

The situation in the Neolithic of the Aeolian Islands is somewhat different, in that it was mainland imports (particularly pottery) which acquired symbolic or prestige importance in island communities. Mainland pottery vessels may have been exchanged for Liparan obsidian but, whilst obsidian seems to have been used in an exclusively utilitarian context, the extent of stylistic investment and the contexts of deposition suggest that south Italian finewares had greater symbolic significance. These finewares, together with obsidian from Lipari and Pantelleria, circulated within a network of contacts which included southern Italy, Sicily, the Aeolian Islands, Malta and Pantelleria (Malone 1985). The presence of both Italian (Scaloria ware) and Dalmatian finewares on the Tremiti Islands hints at a similar (and as yet poorly understood) network of contacts linking communities on either side of the Adriatic Sea. The contrasting character and depositional context of the materials imported to islands such as Lipari and Pantelleria (decorated finewares), and the materials exported from them (obsidian) suggest that, in this case, island/mainland interaction may have been more significant to the social formations of island communities than to those of mainland groups.

Both in the Neolithic of the Aegean and in the Neolithic of the central Mediterranean, the control of interaction between islands and mainlands seems to have been a significant factor in the establishment and reproduction of hierarchical social relations within individual communities (mainland communities in the Aegean, island communities in the central Mediterranean). In both cases, prestige value seems to have been attached to particular items or materials procured through interaction with distant areas, and this value is likely to have depended upon the way in which cultural meaning was assigned to geographical space within the areas known to the communities involved.

Stoddart *et al.* (1993), analysing the evidence from the Maltese Islands, have argued for a cyclical development of 'monument-oriented' and 'exchange-oriented' societies. Monument-oriented societies are those in which social relations are mediated primarily through the construction of monuments and the control of ritual practice: several examples were identified in the previous chapter. Is it possible, then, to identify exchange-oriented societies in Mediterranean island prehistory: societies in which social relations were mediated primarily through the control of inter-island and island/mainland interaction? The evidence is less clear. There is unambiguous evidence for exchange, of course, but to what extent was it the basis of social relations within and between island and mainland communities? In considering this question, we should perhaps bear it in mind that the classic ethnographic example of an exchange-oriented society, the islands of the Melanesian Kula ring, would be difficult to identify as such on archaeological evidence alone, since the items in circulation (shell armlets and necklaces) could not be characterised to source. The evidence for the Neolithic of the Aegean area and the central Mediterranean is certainly compatible with the model of an exchange-oriented society, whilst the circulation of marble figurines in the Aegean Early Bronze Age is even more suggestive. The evidence from the Maltese Islands suggests that island/mainland exchange was least significant during the major phases of monument construction, lending weight to the suggestion of a cyclical alternation of exchange-oriented and monument-oriented societies (Stoddart et al. 1993). To some extent the same pattern can be identified elsewhere. In the Balearic Islands, for example, the development of the monument-oriented Talayotic culture followed the decline of the Beaker package, which should probably be seen as reflecting a prestige exchange system (see Chapter 5). In Sardinia, the monument-oriented Nuraghic culture has been seen as an 'autonomous episode' sandwiched between periods of intense inter-regional interaction (Lewthwaite 1985b).

The situation on Crete is rather different. It has been argued (see Chapter 5) that the palaces of Middle and Late Bronze Age Crete reflect the emergence of a monument-oriented society. As we might expect, the development of the earliest palaces followed the decline of the prestige exchange networks of the Aegean Early Bronze Age. In contrast to the other monument-oriented societies which we have examined, however, Minoan Crete became the centre of a wide-ranging network of international contacts. Contacts within and beyond the Aegean should perhaps be considered separately. Within the Aegean itself, the net flow of resources seems to have been centripetal, from the rest of the Aegean towards Crete. The social significance of this interaction, however, may have been greater in non-Cretan communities, where Minoan imports seem to have had prestige significance (as in the shaft graves of Mycenae), and where Minoan architectural and artistic styles were emulated (as at Akrotiri) as part of the 'Versailles effect' (Wiener 1984). This pattern is typical of a core–periphery situation, which occurs when societies at fundamentally different levels of socio-economic organisation come into contact with one another. The difference in socioeconomic organisation between Middle and Late Bronze Age Crete and other communities in the Aegean area can itself be seen as a function of island sociogeography (see Chapter 5). Once a core–periphery situation had been established in the Aegean, the Minoan elites were able to make strategic use of particular islands (Kythera and the islands of the western and eastern 'strings') to ensure the continued centripetal flow of resources. We see a similar strategic use of islands in the later Mycenean network of contacts with the central Mediterranean, and in Phoenician contacts with Cyprus, the Aegean, Italy and Sardinia. It is difficult to conceive of such

strategic planning other than in the context of a state society, but there
are plenty of comparable examples from more recent history, as witness the
strategic role of islands such as Cyprus, Malta and Gibralter in Medieval
and Post-Medieval times. Contacts between Minoan Crete and areas
beyond the Aegean (most notably Egypt and the Near East) must be seen
in a rather different light since, although Crete constituted a core area with
respect to the rest of the Aegean, it was surely peripheral to the larger-scale
empires of the Near East. Egyptian tomb paintings suggest a significant flow
of goods from the Aegean to Egypt (metals; pottery, probably containing
wine or oil), whilst inscriptions suggest that Cretans also provided services
(as maritime traders and middlemen) to the Egyptian state. In return for
this flow of goods and services, Egyptian and Near-Eastern prestige goods
found their way to Crete. The relationship between Crete and the Near East
is a mirror image of that between Crete and the rest of the Aegean. In both
cases, an advanced state society was able to concentrate its efforts on forming alliances with relatively small island communities, as a means of gaining
access to the resources of a much larger economic hinterland. Just as the
Cretans gained access to the resources of the Greek mainland through their
influence on Kythera, Thera, Melos and Keos, so the Egyptians gained
access to the resources of the Aegean through developing a special relationship with Crete. Although we are dealing here with state societies, we should
not lose sight of Helms' (1988) arguments regarding the links between
geographical distance and social power. The emergence of a state society in
Middle Bronze Age Crete, for example, may have enhanced the cultural
significance of contact with that island: Classical mythology is rich with
references to King Minos, the Minotaur and human sacrifice, suggesting
that Crete was seen by other Aegean communities as a centre of wealth
and power, but also as a place of mystery and danger. Prestige goods from
Crete, including jewellery and weapons, may have been considered to be
imbued with some of these qualities. Egypt and the Near East may have
occupied a similar role in the cosmology of Cretan communities.

Crete and Cyprus are unique among the islands of the Mediterranean, in
having supported long-lived urban and literate civilisations during the
Bronze Age. This is probably due to a combination of circumstances. Both
are among the largest islands in the Mediterranean, and have relatively
diverse ecosystems, permitting significant surplus production. In the case of
Crete, this surplus seems to have been mobilised by an elite whose power
was based upon control of religion and ceremonial practice. The storage
facilities in the Minoan palaces (particularly those of the First Palace
period) suggest that the extent of surplus accumulation was very much greater
than in the other monument-oriented societies which we have identified
(see Chapter 5). Both Crete and Cyprus are also strategically located as
'stepping stones' between mainland areas, and this was a significant factor
in attracting Near Eastern influences. The role of these influences should

not be underestimated: the introduction of writing in Crete and Cyprus was undoubtedly inspired by Near Eastern scripts, and even the form of the Minoan palaces has Near Eastern parallels. The role of Near Eastern contacts was naturally greater in the case of Cyprus, which occupied a strategic position between the civilisations of Egypt, the Levant and southeast Anatolia. Of equal importance in the case of Crete was the development of a network of contacts within the Aegean area itself, allowing Minoan elites to mobilise resources from a large economic hinterland. This was made possible both by the geographical position of Crete, as the central link in a chain of stepping-stone islands stretching between Europe and western Asia, but also by the way in which Crete and its civilisation were viewed by other Aegean communities. The unique position of Crete in the Middle and Late Bronze Age is a theme to which we shall return in the final chapter.

7

ISLAND SOCIOGEOGRAPHY AND MEDITERRANEAN PREHISTORY

This book set out to explore the effects of insularity on the development of human societies, through a range of case studies taken from Mediterranean prehistory. Central to this aim was an attempt to develop a body of theory which moves beyond the framework of the biogeographical approaches which have dominated the literature on island archaeology over the past decade. This is not to deny the value of models based on island biogeography theory, but rather to build on such models by exploring the differences, as well as the similarities, between island populations of humans and other species. The 'theory of island biogeography', as developed by MacArthur and Wilson (1967) was not intended simply as a statement on island populations: rather it set out to use island studies as the basis for understanding more general evolutionary and ecological principles, following on from the pioneering work of Charles Darwin (1968 [1859]). Whilst some of the premises of island biogeography theory might now be questioned (Williamson 1981), its useful- ness is demonstrated by the role that it has played in the design and plan- ning of nature reserves, which are themselves 'islands' in an ecological sense, even if they are completely landlocked. In much the same way, this book set out to explore Evans' (1973) suggestion that islands can be seen as 'labora- tories' for the study of sociocultural processes. In developing these approaches, the aim has been to bridge the gulf between scientific and humanistic approaches to archaeology, by employing the methods of the latter to look at questions which have normally been considered within the domain of the former. In this way it may be possible both to make science more directly relevant to the study of human society, and to make the methodology of humanistic study more rigorous.

The case studies examined in this book provide a basis for the discussion of three types of relationship: the relationship between human communities and the natural environment, relationships between individuals within a community and relationships between communities.

HUMAN COMMUNITIES AND THE NATURAL ENVIRONMENT

Since much of the published literature on island archaeology makes use of models borrowed from ecology and evolutionary biology, one might expect island studies to tell us a great deal about the relationships between human communities and the natural environment (see Chapter 4). In an age in which biodiversity has become a major political issue, the understanding of these relationships is perhaps one of the most important contributions science can make to the future development of human society. This link between archaeology and public concern for the environment has been siezed upon by Bahn and Flenley (1992), in a study which uses resource depletion on Easter Island as a model for the human depletion of the Earth's resources. Easter Island, however, is a unique case, in that it is a virtually closed system, so remote from other land masses as to provide a plausible model for the totally closed system of the Earth itself. Such is not the case with any of the Mediterranean islands. MacArthur and Wilson (1967) develop a series of ecological predictions on the basis of island studies of animal and plant species. Fundamental to their model is the suggestion that island ecosystems are characterised by reduced biodiversity, a factor to which colonising species must adapt if they are to survive and reproduce. Whilst this is undoubtedly true of small, remote islands such as Easter Island, the effect of reduced biodiversity on the islands of the Mediterranean seems to have been limited. Larger islands, which according to island biogeography theory should have a greater degree of biodiversity, do seem, on the whole, to have been colonised earlier than smaller islands, and size (or biodiversity) seems to have been a more important factor than remoteness in determining the likelihood of colonisation (see Chapter 3). It seems clear, however, that for human communities (unlike other species), island ecosystems had advantages as well as disadvantages. Humans are unique among animals in having an 'extra-somatic means of adaptation' in cultural behaviour, especially tool use: biologically, they are also more omnivorous than most animal species. Cultural adaptations have the advantage that they can take place far more rapidly than biological adaptations, since they are not dependent upon the chance factor of genetic mutations. The ability of humans to build boats and make tools allowed them to exploit marine resources to a far greater extent than any other terrestrial mammal: since marine resources are not subject to the 'island effect' this allowed them to overcome the effect of reduced biodiversity, and to exploit small islands such as Paros, Mykonos and Antiparos (Evans and Renfrew 1968; Bintliff 1977), which certainly could not support viable populations of other large primates. The use of boats also meant that for humans, unlike other species, islands were not necessarily closed or even semi-closed systems. In some cases, as in the Early Neolithic of Crete and Cyprus (Broodbank and Strasser 1991), human colonists established themselves on

180

islands with ready-made ecosystems comprising domesticated crops and animals. Any attempt to understand the relationship between human communities and island ecosystems must take account of the unique adaptive strategies of *Homo culturalis* and, for this reason, the island biogeography model of MacArthur and Wilson (1967), whilst it may provide important insights, will always be limited in its applicability. The ability of human communities to adapt rapidly to new ecological conditions is clearly demonstrated by the range of specialised economies developed on the Mediterranean islands during the Neolithic: the exploitation of migratory fish shoals in the Saliagos Group (Evans and Renfrew 1968; Bintliff 1977), the selective introduction of domestic animals to Corsica and Sardinia (Vigne 1987; Sanges 1987), the corralling of endemic *Myotragus* on the Balearic Islands (Waldren 1982).

MacArthur's and Wilson's (1967) concept of 'species equilibrium' suggests that the successful establishment of any new species in an island ecosystem will inevitably result in the extinction of previously established species. Following Bahn and Flenley (1992), one might expect the human colonisation of islands to have resulted in severe resource depletion and reduced biodiversity. Whilst the 'blitzkrieg' (cf. Mosimann and Martin 1975) model of species extinction seems to be applicable in the case of pygmy elephant and hippopotamus on Cyprus (Simmons 1991), it is certainly not applicable in the case of *Prolagus*, which survived for at least 4000 years alongside human communities in Corsica and Sardinia, or in the case of *Myotragus* in the Balearic Islands, which coexisted with humans for around 2800 years. The concept of 'species equilibrium' is an oversimplification, since the probability of extinction depends fundamentally upon the reproductive biology of the individual species concerned, and on the nature of the ecological relationship between them (Williamson 1981). In the case of both *Prolagus* and *Myotragus*, the introduction of domestic crops and livestock, and the accompanying destruction of habitat, seems to have had a far more dramatic effect than direct human predation.

Whilst political commentators on the human threat to the environment have tended to take the industrial revolution as a starting point, it may well be that the transition to food production was a more significant juncture in the relationship between humans and the global ecosystem. This brings us to a further difference between humans and other animals. Most animals have a relatively stable relationship with their environment. Their consumption of resources is directly related to demography: as population rises, so consumption of resources increases, until such time as the environment can no longer support this, and the population crashes. Despite short-term cyclical fluctuations, therefore, population and resource consumption remain relatively stable in the long term. This is not the case with humans: cultural adaptations have allowed them, at various points in time, to fundamentally change their relationship to the natural environment, and to dramatically increase the consumption of resources. The transition to food production is the most dramatic example of such an adaptation in prehistory, the 'secondary-products revolution' (Sherratt 1979) is perhaps another. These adaptations are not necessarily linked to increases in human population since humans, unlike other species, may accumulate a food surplus for social and political reasons: *Homo culturalis* is also, fundamentally, *Homo socialis*. These cultural adaptations, however, do have the effect of increasing the 'carrying capacity' of land masses (cf. Williamson and Sabath 1984), and the effect of this can be clearly seen in the pattern of island colonisation in Mediterranean prehistory (see Chapter 3): three distinct phases of island colonisation (Figure 3.10) can be distinguished, and these coincide closely with the transition to food production, the 'secondary-products revolution' and the emergence of urban civilisations. The process of surplus production and appropriation in human societies tends to be an expansionary one (Patton 1993), which places pressure both on the natural environment and on the social formation, leading to a cyclical pattern of societal development and collapse. Such cycles can clearly be identified in Mediterranean prehistory, for example, in the rise and fall of the Minoan 'palace civilisation'. Interestingly, however, the collapse of social systems has rarely (if ever) resulted in any long-term shift to less intensive systems of production, so that the effect on the environment is cumulative rather than cyclical and, as a result of this, the 'wavelength' of long-term socioeconomic cycles has tended to decrease with time. Like Bahn's and Flenley's (1992) study of Easter Island, this has interesting (if disturbing) implications for modern society. It is difficult for any politician, of whatever persuasion and in whatever country, to accept the inevitable conclusion that continued economic growth in the long term is unsustainable, but this may be the greatest challenge for the twenty-first century. Certainly we should not imagine ourselves to be immune from the effects of the long-term processes which can be identified in human history and prehistory: whether, having understood these processes, we can actually change them, is something that time alone will tell.

The case studies outlined in this book (see especially Chapters 3 to 4) show a range of human cultural adaptations to distinctive island environments, some of them successful (at least in the medium term), others less so. Island biogeography theory, though helpful in some cases, cannot adequately explain the complexity of these adaptations. The limitations of the biogeographic approach, however, may themselves prove useful in understanding the unique ecological position of the human animal, and in attempting to elucidate the relationships between ecology and society, culture and nature. This may have implications which go far beyond the study of prehistoric societies on the Mediterranean islands.

INTRA-COMMUNAL SOCIAL RELATIONSHIPS

Ethnographic studies of 'tribal' and 'chiefdom' societies suggest that social power may operate, on the one hand, through the elite control of sacred knowledge and ritual practice, often linked to initiation ceremonies (Meillassoux 1972), and on the other hand through the control of access to socially valued material goods (Bender 1985). In both instances, the elite (generally lineage elders or a chiefly caste) is able to make demands upon the labour of young people by controlling their route to economic independence and to physical and social reproduction. These forms of social power may operate simultaneously in a given society and may, in ideological terms, be inextricably woven together (Patton 1993). In an island context, it is not entirely surprising to find that one or other of these approaches to social power may be emphasised to the exclusion of the other. Thus we can distinguish 'monument-oriented' societies, such as Malekula (Deacon 1934; Layard 1942), in which power operates through the control of elaborate megalithic ritual, and 'exchange-oriented' societies, such as the Melanesian Kula ring (Malinowski 1922), in which power is based on the control of ritualised exchange (see Chapter 2). In the context of Mediterranean prehistory, the Talayotic monuments of the Balearic Islands, the stone temples of Malta and the palaces of Minoan Crete can perhaps be seen as indications of monument-oriented societies (see Chapter 5), whilst exchange-oriented societies can be identified in the Neolithic of the Aegean and the central Mediterranean (see Chapter 6).

Monument-oriented societies on islands typically show a high degree of cultural uniqueness: the material culture, architecture and religion of certain islands at particular moments in time is distinctly different from that of adjacent mainlands and neighbouring islands. In the case of remote oceanic islands, such as Easter Island, this is likely to result from a genuine lack of contact with the outside world. This is not a likely scenario on the Mediterranean islands, however, where the builders of, for example, the Maltese temples and the Minoan palaces, seem to have deliberately emphasised the unique cultural identity of the island communities to which they belonged. The monumental traditions examined in Chapter 5 all show tendencies towards increasing elaboration through time: as part of this process, the monuments themselves become larger, and their structure more complex, suggesting increasing restriction of access. This is perhaps most clearly seen in the case of Neolithic Malta, where the relatively small and simple temples of the Ggantija phase (themselves probably derived from simple shrines such as that found at Skorba) were replaced in the Tarxien phase by much larger constructions with a more complex use of space (Stoddart *et al.* 1993): areas which had been open and visible (e.g. the central apse at Skorba) were sealed in and hidden by internal walls, and the most important concentrations of cult objects and images are increasingly found in these innermost recesses. A similar process of elaboration can be seen in the case of Minoan Crete, where the palaces of the Late Bronze Age are significantly larger than those of the Middle Bronze Age, and where iconographic evidence suggests that ceremonial activities may also have become increasingly complex. These processes of elaboration are by no means unique to island communities: a similar process was identified, for example, in a study of Neolithic communities in Brittany (Patton 1993), where megalithic monuments became increasingly large through time, and where access to the monuments seems also to have become increasingly restricted. In attempting to develop a model to explain this process of monumental elaboration in Brittany, it was suggested (Patton *op. cit.*) that the social formations of Neolithic communities embodied structural tendencies towards expansion, intensifying competition and increasing concentration of power. These trends, as Friedman and Rowlands (1977) stress, require the production and appropriation of increasingly large surpluses, putting pressure both on the natural environment and on the social formation, and leading ultimately to crisis and collapse. Such a model would be equally applicable to the case studies examined in Chapter 5.

In an exchange-oriented society, it is a community's links with the outside world, rather than its unique cultural identity, that is emphasised in material culture and social practice. Social power in such communities operates primarily through elite control over the circulation of exotic items, often required for socially significant transactions such as bridewealth payments. These items generally carry prestige value, and may be considered to have sacred or magical properties: archaeologically, we might expect this significance to be reflected in the formal deposition of imported objects, for example, in graves and votive hoards. In an island situation, the control of navigational knowledge and access to boats may provide an additional mechanism for the elite control of socially significant exchange, as in the case of Melanesian Kula (Malinowski 1922; Munn 1986). Once again, however, the island context may simply be giving us a clearer view of much more general cultural processes and phenomena. Helms (1988), taking a cross-cultural perspective and looking at both island and mainland societies, has drawn attention to the 'widespread association of political elites with foreign and distant goods and information', suggesting that space and distance are commonly accorded political and ideological significance, and that geographical distance is frequently equated with supernatural distance. In the mainland case-study of Neolithic Brittany already alluded to (Patton 1993) exchange (of stone axes) seems to have had a fundamental social and symbolic importance, intimately linked to the meaning and social context of megalithic monuments. The island case studies examined in this book show a greater degree of polarisation between monument-oriented and exchange-oriented strategies for the establishment and reproduction of power relationships: the communities of Neolithic Brittany cannot be characterised as either monument-oriented or exchange-oriented, since the two strategies were interwoven with one another as part of a single ideology. The 'island effect' allows us to separate these two components of ideology and social relations, and to develop a better understanding of the dynamics which underlie them.

Control of island/mainland interaction may be significant to the social formations of both island and mainland societies. In the Neolithic and Early Bronze Age of the Aegean, for example, it was objects from the Cycladic islands (stone vases and figurines) which acquired social significance in the communities of the Greek mainland (Perlès 1979), whereas in the Neolithic of the central Mediterranean, pottery from the Italian mainland became significant to the communities of the Aeolian and Maltese Islands (Malone 1985). In both cases, however, the control of interaction *between* communities seems to have been a significant factor in the reproduction of social relations *within* communities. Social and cultural significance need not apply equally to all items and materials circulating within an exchange network: in the Aegean, for example, the exchange of obsidian seems to have been undertaken on an essentially utilitarian basis, in contrast to the circulation of marble vases and figurines (Perlès 1979; Torrence 1986). Even in the Trobriand Islands (Malinowski 1922) inter-island relations involve both 'Kula' (ritualised exchange of valuables) and 'Gimwali' (barter of foodstuffs and other items), though these are kept separate and operate according to different rules. Like the monument-oriented societies already discussed, exchange-oriented societies may show tendencies towards expansion, intensifying competition and increasing concentration of power. This is most clearly seen in the Neolithic and Early Bronze Age of the Aegean, where the exchange of Cycladic marble objects begins on a relatively small scale in the Neolithic, and increases dramatically in Early Cycladic I–II (2700–2300 cal. BC). The objects themselves become more elaborate (the most complex 'musician' figurines and composite groups occurring in ECII) and clear concentrations of figurines can be identified on particular sites (such as Aghious Kosmas on the Saronic Gulf, Aghia Fotia and Archanes on Crete) outside the Cyclades. As with the 'Monument Oriented' societies, these expansionary trends may put pressure on both the environment and the social formation of the communities concerned, leading to crisis and ultimate collapse. The emergence of the monument-oriented palace civilisation of Minoan Crete in the Middle Bronze Age was preceded by the collapse of the exchange-oriented system which dominated the Early Bronze Age of the Aegean region.

The pattern of cyclical alternation of monument-oriented and exchange-oriented systems noted by Stoddart *et al.* (1993) in the Neolithic of the Maltese Islands finds parallels in the prehistory of the Aegean, Sardinia and Corsica and the Balearic Islands (see Chapter 5). This pattern may provide further insights into the nature of social dynamics in tribal societies. In a previous publication (Patton 1993), a cyclical model of social dynamics was suggested, involving three main phases, characterised as follows:

1 *Consolidation.*
  This phase is marked by increasing social differentiation and centralisation, requiring increasing surplus production. In the case of 'monument-oriented'

societies, monuments become larger, whilst access to them is increasingly restricted. In the case of 'exchange-oriented' systems, interaction becomes increasingly competitive and the exchanged valuables increasingly elaborate.

2 *Crisis*.

Surplus production and appropriation are constrained by environmental, technological and social factors and, at a given point, these constraints produce a crisis in the social formation. Typically the elite attempts to adapt to this crisis by changing its legitimation strategy, de-emphasising social differentiation in an attempt to minimise the crisis.

3 *Replacement*.

If the crisis continues, it is likely to lead to the replacement of one elite structure by another. This phase may be reflected in the archaeological record by fundamental changes in material culture and social practice. In the case studies outlined in this book, this is the phase during which a monument-oriented system is likely to be replaced by an exchange-oriented system, or *vice versa*. The phase is generally marked by decentralisation and the collapse of social hierarchy, but this is immediately followed by a new phase of centralisation and increasing social differentiation as a new cycle begins.

This model implies the existence of two competing power structures during the 'crisis' and 'replacement' phases which mark the transition between one cycle and another. New elite structures may have challenged the power of traditional institutions by establishing patron/client relationships which offered 'clients' an alternative route to economic independence and social reproduction. This would further limit the ability of the traditional elite to requisition surplus production, thus hastening its collapse. Dominance would then pass to the new elite structure at the beginning of a new 'consolidation' phase.

The polarisation of monument-oriented and exchange-oriented systems in the island case studies makes these processes more visible, and more easy to comprehend. The processes are not specific to island communities, but these communities do provide ideal conditions for their study, in much the same way as the Galapagos Islands provided Darwin with the evidence he needed to develop his general theories of biological evolution.

The case of Minoan Crete deserves some further comment, since in two respects it is unique among the case studies which we have examined in detail (Bronze Age Cyprus is in some respects comparable, but largely because of its proximity to the urban civilisations of Egypt and the Near East). First, Minoan Crete is unique in that it supported a long-lived civilisation which was both urban and literate, and which was characterised by a much higher degree of social differentiation than is in evidence in the other societies discussed in this book. Second, Minoan Crete is unique among the monument-oriented societies which we have identified, in the extent to which the

Minoan elites also engaged in international trade. In attempting to identify the factors which encouraged the development of Europe's first urban civilisation on the island of Crete, it is clear that intra-communal and inter-communal relations are inextricably linked to one another.

INTER-COMMUNAL SOCIAL RELATIONS

We have already touched upon the subject of relationships between communities, insofar as they are significant to the establishment and reproduction of social relations within communities. There are, however, other dimensions to inter-island and island/mainland relationships which may provide further insights into the nature and significance of exchange and contacts between communities.

Many of the social interpretations which have been developed or adapted by archaeologists over the past decades have been based upon an implicit model of exclusive territorial control; the idea that human communities (with the possible exception of hunter–gatherer groups) typically live in clearly defined territories, within which they have exclusive control over resources. In contrast to this, the Neolithic and Early Bronze Age evidence, both from the Aegean (Torrence 1986) and from Sardinia (Tykot 1992) suggests that mainland groups may have had direct access to island sources of obsidian (see Chapter 6). In the case of the Aegean this is perhaps less surprising since, at least in the initial stages of obsidian exploitation, the island of Melos (the main source of obsidian in the Aegean) was apparently unoccupied. Torrence's (1986) research, however, suggests that mainland communities continued to have direct access to the Melian obsidian sources throughout the Early Bronze Age, when the island clearly did support a permanent population, whilst in the case of Sardinia, the island was occupied throughout the period in question. Whilst this 'direct access' scenario casts doubt on the whole assumption of exclusive territorial control, there are several reasons for suggesting that it is less unlikely than it may at first appear. For one thing, there is the sheer abundance of the obsidian sources on Melos and Sardinia: several millennia of exploitation by both mainland and island communities did not even come close to exhausting them. To allow outside groups direct access to these resources, therefore, would cost the island community very little, and there was no need to add labour or value to what was already a very valuable resource. The island communities, in turn, may have been dependent upon mainland groups for access to commodities and materials (not necessarily visible in the archaeological record) essential to their physical survival or social reproduction. Close alliances with mainland groups may have been critically important to small island communities, particularly in the Aegean: such alliances could act as a form of insurance policy against the possibilities of resource failure and famine, which would be particularly dangerous in the early stages of island colonisation. Set against the benefits to island communities of such alliances, allowing mainland groups free access to particular resources, such as obsidian outcrops, may have seemed a very small price to pay. In the case of the Aegean Early Bronze Age, it seems eminently possible that island/mainland alliances were based upon kinship, and that colonising groups maintained close bonds with their parent populations on the mainland over many generations. Mainland groups may then have continued to exercise certain rights over the island's resources, just as they had prior to colonisation. The bonds between an island community and its parent group on the mainland may have been strengthened through regular exchange of marriage partners, something which would in any case be necessary if a small colonising population was to reproduce itself without incest.

Whilst the relationships between mainland and island communities in the Aegean Early Bronze Age may have been essentially reciprocal, the same cannot be said for the relationships which developed between Crete and other island and mainland groups in the Middle and Late Bronze Age. The presence of over forty Cycladic amphorae and storage jars in Middle Minoan III contexts at Knossos (MacGillivray 1984) suggests that at least some Cycladic communities had become incorporated in the redistributive economy organised through the Cretan palaces. This evidence may signal the emergence of a core/periphery situation within the Aegean region (see Chapter 6). Core/periphery relationships typically have the following characteristics, all of which can be identified in the relationship between Crete and other Aegean communities in the Middle and Late Bronze Age.

1 Core/periphery relationships generally develop between societies with very different social formations. In most cases the social formation of the core area is characterised by a much greater degree of centralisation and social differentiation than that of peripheral communities.
2 The net flow of resources within a core/periphery relationship is always centripetal, from periphery to core.
3 Core/periphery relationships may result in increasing competition and social differentiation within the peripheral communities. This is often articulated through the emulation of the core society (Wiener's (1984) 'Versailles effect'), and through the possession of prestige goods from the core area (as witness, for example, the importance of Cretan prestige goods in the shaft graves of Mycenae).

The emergence of a monument-oriented palace civilisation on the island of Crete in the Middle Bronze Age created the regional 'differentials', on the basis of which a series of core/periphery relationships could be established: by the beginning of Middle Minoan III, the social formation of Crete was fundamentally different from that of other communities in the Aegean. By fostering the development of core/periphery relationships, Minoan elites were able to mobilise resources from a greatly expanded hinterland, allowing further economic growth without initially provoking the kind of environmental and social pressures which checked the expansion of monument-oriented systems in the other case studies examined in this book. The distribution of Minoan artefacts and 'Minoanised' sites within the Aegean region suggests that these relationships were established as part of a deliberate policy on the part of the Minoan elites. By forming close alliances with a relatively small number of strategically placed island communities (Kythera and the islands of the western and eastern 'strings'), Crete was able to ensure a centripetal flow of resources from all parts of the Aegean (Niemeier 1984 and see Figure 6.7). Within such a system, each of the 'Minoanised' islands would act, in a sense, as a subsidiary core, passing Minoan prestige goods on to a wider hinterland in return for larger quantities of metals, foodstuffs and other commodities. We can perhaps see this as a model for the development of core/periphery relationships elsewhere. In the sixth century BC, for example, the Greek trading colonies of the western Mediterranean seem to have established alliances with a small number of elite groups, strategically positioned along the Rhone Valley, a natural artery of communication between the Mediterranean and the central European hinterland. Through this network of alliances the Greeks were able to ensure a centripetal flow of resources from central European periphery to Mediterranean core. Although in this case we are dealing with relationships between mainland groups, rather than with inter-island or island/mainland interaction, the processes involved are remarkably similar to those observed in the Late Bronze Age of the Aegean. These processes, however, are more clearly visible and more easily understood in the Aegean case-study, where the contrast between the 'Minoanised' and 'non-Minoanised' islands cries out for explanation.

One of the features of core/periphery relationships is that they tend to result in rapid social and economic change within the peripheral communities. This process is clearly seen in the Aegean Late Bronze Age, with the emergence of the Mycenean palace civilisation on the Greek mainland. As the Late Bronze Age progressed, mainland communities became increasingly hierarchical, increasingly centralised and increasingly powerful. The 'differential' between Cretan and mainland communities broke down, and the core/periphery relationship between them collapsed. The Minoan elites had depended for their power and wealth upon a centripetal flow of resources from around the Aegean, which increasingly dried up. The social reproduction of these elites, with their elaborate palaces and their entourage of craft and religious specialists, required the mobilisation of a surplus vastly in excess of that which could be produced in Crete itself. The progressive 'de-peripheralisation' of its economic hinterland, therefore, inevitably brought about the collapse of the Minoan palace civilisation itself. We could doubtless draw parallels with other instances of state collapse, such as the fall of the Roman Empire or, more recently, the disintegration of the Soviet bloc. The processes involved are certainly not specific to island communities but, once again, they are particularly visible in the island context.

THE 'ISLAND LABORATORY' REVISITED

When John Evans (1973) suggested that islands could be seen as 'laboratories' for the study of sociocultural processes in the human past, he doubtless had in mind the enormous progress that has been made in ecology and evolutionary biology on the basis of island studies, beginning with Darwin's (1968 [1859]) research on the Galapagos Islands and elsewhere. The study of island populations continues to be an important strand of ecological and evolutionary research (MacArthur and Wilson 1967; Williamson 1981) and has made considerable contributions to the development of a body of general theory within the broader disciplines of zoology and botany. Evans' (1973) suggestion was that island studies could make a similar contribution to the development of general theory within archaeology (see Chapter 1). Most of the attempts that have been made to use island studies in this respect (Cherry 1981, 1990; Terrell 1986) have been grounded explicitly in the theoretical and methodological framework of 'island biogeography' (MacArthur and Wilson 1967). Although these approaches have provided important insights, this book began with the contention that it was necessary to move beyond them, and to develop a body of theory specific to the study of human island populations. This book represents a preliminary attempt at developing such a body of theory, and has perhaps gone some way towards demonstrating the potential of island studies in understanding the dynamics of human societies. Studies based upon island biogeography theory have tended, not surprisingly, to focus on the dynamics which underlie the relationships between human communities and their environment. What I hope to have shown in this book is that island studies can make an equally important contribution to understanding the dynamics of intra-communal and inter-communal social relations. There is no simple dichotomy, however, between studies of human ecology and studies of human society: ecological and social relationships are linked by a complex web of interconnections. In studying the development of human island populations, therefore, it is necessary neither to ignore, nor indeed to reject island biogeography theory, but rather to adapt it and to reincorporate it within a larger body of theory which is concerned with all aspects of the dynamics of human society.

BIBLIOGRAPHY

Ammerman, A.J. (1985) *The Acconia Survey: Neolithic Settlement and the Obsidian Trade.* London, Institute of Archaeology Occasional Paper no. 10.

Ammerman, A.J. and Andrefsky, W. (1982) 'Reduction Sequences and the Exchange of Obsidian in Neolithic Calabria'. In Ericson and Earle (eds), 149–171.

Anastassiades, P. (1949) 'General Features of the Soils of Greece'. *Soil Science* 67, 347–362.

Arca, M., Martini, F., Pitzalis, G., Tuveri, C. and Ulzega, A. (1982) *Il Paleolitico dell'Anglona (Sardegna Settentrionale).* (Quaderni 12), Sassari, Dessi.

Atzeni, E. (1980) 'Menhirs Anthropomorphi e Statue-Menhirs della Sardegna'. *Annali del Museo Civico della Spezia* 2, 9–40.

Atzeni, C., Massidda, L., Sanna, V. and Virdis, P. (1987) 'Archeometallurgia Nuragica nel Terrioria di Villanovaforru'. In G. Lilliu, G. Ugas and G. Lai (eds) *La Sardegna nel Mediterraneo tra il Secondo e il Primo Millennio a.c.* Cagliari, Amministrazione Provinciale di Cagliari, 147–165.

Bahn, P. and Flenley, J.R. (1992) *Easter Island, Earth Island.* London, Thames & Hudson.

Balmuth, M.S. (1984) 'The Nuraghi of Sardinia: An Introduction'. In Balmuth and Rowland (eds), 23–52.

Balmuth, M.S. (ed.) (1987) *Studies in Sardinian Archaeology III. Nuraghic Sardinia and the Mycenean World.* Oxford, British Archaeological Reports (International Series) no. 387.

Balmuth, M.S. (1992) 'Archaeology in Sardinia'. *American Journal of Archaeology* 96, 663–697.

Balmuth, M.S. and Rowland, R.J. (eds) (1984) *Studies in Sardinian Archaeology.* University of Michigan Press.

Barber, R.L.N. (1984) 'The Status of Phylakopi in Creto-Cycladic Relations'. In Hägg and Marinatos (eds), 179–182.

Barber, R.L.N. (1987) *The Cyclades in the Bronze Age.* London, Duckworth.

Bass, G.F. (1967) *Cape Gelidonya: A Bronze Age Shipwreck.* Philadelphia, American Philosophical Society.

Bass, G.F. (1986) 'The Bronze Age Shipwreck at Ulu Burun (Kas): 1984 Campaign'. *American Journal of Archaeology* 90, 269–296.

Bass, G.F. (1991) 'Evidence of Trade from Bronze Age Shipwrecks'. In Gale (ed.), 69–82.

Bates, M. (1963) 'Nature's Effect and Control of Man'. In Fosberg (ed.), 101–116.

Bender, B. (1978) 'Gatherer-Hunter to Farmer: A Social Perspective'. *World Archaeology* 10, 204–222.

Bender, B. (1985) 'Emergent Tribal Formations in the American Mid-Continent'. *American Antiquity* 50 (1), 52–62.

Benzi, M. (1984) 'Evidence for a Middle Minoan Settlement on the Acropolis at Ialysos (Mount Philerimos)'. In Hägg and Marinatos (eds), 93–105.

Bernabo Brea, L. and Cavalier, M. (1960) *Meligunis Lipara. vol. I.* Palermo, Flaccovio.

Bernabo Brea, L. and Cavalier, M. (1980) *Meligunis Lipara. vol. IV.* Palermo, Flaccovio.

Berry, R.J. (1979) 'The Outer Hebrides: Where Genes and Geography Meet'. *Proceedings of the Royal Society of Edinburgh* 77, 21–43.

Bianchini, G. (1969) 'Risultati delle Richerce sul Paleolitico Inferiore in Sicilia e la Scoperta di Industrie del Gruppo della <Pebble Culture> nei Terrazzi Quaternari di Capo Rosello in Territorio di Realmonte'. *Atti della XIII Riunione Scient. del'I.P.P. Siracusa-Malta 1968*, 89–109.

Bietti-Sestieri, A.-M. (1985) 'Contact, Exchange and Conflict in the Italian Bronze Age: The Myceneans on the Tyrrhenian Coasts and Islands'. In Malone and Stoddart (eds), 305–337.

Binford, L.R. (1962) 'Archaeology as Anthropology'. *American Antiquity* 28, 217–225.

Binford, L.R. (1965) 'Archaeological Systematics and the Study of Culture Process'. *American Antiquity* 31, 203–210.

Bintliff, J. (1977) *Natural Environment and Prehistoric Settlement in Prehistoric Greece.* Oxford, British Archaeological Reports (Supplementary Series) no. 28.

Blanc, A.C. (1955) 'Notizie Preliminari sull'Attività Scientifica dell'Istituto Italiano di Paleontologia Umana nel 1955'. *Quaternaria* 2, 310–311.

Blance, B. (1961) 'Early Bronze Age Colonists in Iberia'. *Antiquity* 35, 192–202.

Bonnanno, A., Gouder, T., Malone, C. and Stoddart, S. (1990) 'Monuments in an Island Society: The Maltese Context'. *World Archaeology* 22, 190–205.

Bosanquet, R.C. and Mackenzie, D. (1904) *Excavations at Phylakopi in Melos.* London, Macmillan.

Bossert, E.M. (1967) 'Kastri auf Seros'. *Archaiologikon Deltion* 22, 53–76.

Bradley, R. (1990) *The Passage of Arms: An Archaeological Analysis of Prehistoric Hoards and Votive Deposits.* Cambridge University Press.

Branigan, K. (1970) *The Tombs of Mesara.* London, Duckworth.

Branigan, K. (1981) 'Minoan Colonialism'. *Annual of the British School of Archaeology at Athens* 76, 23–33.

Branigan, K. (1984) 'Minoan Community Colonies in the Aegean'. In Hägg and Marinatos (eds), 49–52.

Broodbank, C. (1992) 'The Neolithic Labyrinth: Social Change at Knossos before the Bronze Age'. *Journal of Mediterranean Archaeology* 5, 39–75.

Broodbank, C. and Strasser, T.F. (1991) 'Migrant Farmers and the Neolithic Colonisation of Crete'. *Antiquity* 65, 233–245.

Brown, J.H. and Kodric-Brown, A. (1977) 'Turnover Rates in Insular Biogeography: Effects of Immigration on Extinction'. *Ecology* 58, 445–449.

Burkert, W. (1993) 'Concordia Discors: The Literary and Archaeological Evidence on the Sanctuary of Samothrace'. In N. Marinatos and R. Hägg (eds) *Greek Sanctuaries: New Approaches.* London, Routledge, 178–191.

Cadogan, G. (1987) 'What Happened at the Old Palace of Knossos?' In Hägg and Marinatos (eds), 71–74.

Campbell, S. (1983) 'Kula in Vakuta: The Mechanics of Keda'. In Leach and Leach (eds), 201–227.

Camps, G. (1976) 'Navigations et Relations Interméditérraneennes Préhistoriques'. *Actes du ixè Congrès International des Sciences Préhistoriques et Protohistoriques,* Nice.

192

Camps, G. (1981) *Terrina IV (Aleria-Haute, Corse). Campagne de Fouilles de 1981.* Travaux du Laboratoire d'Anthropologie des Pays de la Méditerranée Occidentale.

Canal, J. and Carbonell, E. (1979) *Las Estaciones Préhistoriques del Puig d'En Roca.* Girona, Edisione del Assosiazione Arqueologica de Girona.

Castaldi, E. (1984) 'Cultura Calcolitica di Monte Claro nel Sito de Biriai (Oliena-Nuoro, Sardegna)'. In Waldren *et al.* (eds), 567–590.

Catling, H. (1986) 'Archaeology in Greece 1985–6'. *Journal of Hellenic Studies* 32, 3–101.

Catling, H. and MacGillivray, J.A. (1983) 'An Early Cypriot III Vase from the Palace at Knossos'. *Annual of the British School of Archaeology at Athens* 78, 1–8.

Cherry, J.F. (1978) 'Generalisation and the Archaeology of the State'. In D. Green, C. Haselgrove and M. Spriggs (eds) *Social Organisation and Settlement.* British Archaeological Reports (International Series) 47, 411–437.

Cherry, J.F. (1981) 'Pattern and Process in the Earliest Colonisation of the Mediterranean Islands'. *Proceedings of the Prehistoric Society* 47, 41–68.

Cherry, J.F. (1984) 'The Initial Colonisation of the West Mediterranean Islands in the Light of Island Biogeography and Palaeogeography'. In Waldren *et al.* (eds), 35–67.

Cherry, J.F. (1990) 'The First Colonisation of the Mediterranean Islands: A Review of Recent Research'. *Journal of Mediterranean Archaeology* 3, 145–221.

Cherry, J.F. (1992) 'Palaeolithic Sardinians? Some Questions of Evidence and Method'. In Tykot and Andrews (eds), 43–56.

Chrysolaki, S. and Platon, L. (1987) 'Relations Between the Town and Palace of Zakros'. In Hägg and Marinatos (eds), 77–84.

Clutton-Brock, J. (1984) 'Preliminary Report on the Animal Remains from Ferrandell-Oleza, with comments on the extinction of *Myotragus balearicus*, and on the introduction of domestic livestock to Mallorca'. In Waldren *et al.* (eds), 99–118.

Coldstream, J.N. (1988) 'Early Greek Pottery in Tyre and Cyprus'. *Report of the Department of Antiquities of Cyprus 1988*, 43.

Coldstream, J.N. and Huxley, G.L. (1984) 'The Minoans of Kythera'. In Hägg and Marinatos (eds), 107–112.

Coles, J.M. and Harding, A.F. (1979) *The Bronze Age in Europe.* London, Methuen.

Constantini, L., Piperno, M. and Tusa, S. (1987) 'La Néolithisation de la Sicile Occidentale, d'Aprés les Resultats des fouilles à la Grotte d'Uzzo (Trapani)'. In Guilaine *et al.* (eds), 397–405.

Contu, E. (1984) 'Monte d'Accodi, Sassari: Problematiche di Studio e di Ricerca di un Singulare Monumento Preistorico'. In Waldren *et al.* (eds), 591–610.

Darwin, C. (1968 [1859]) *The Origin of Species by Means of Natural Selection, or the Preservation of Favoured Races in the Struggle for Life.* Oxford University Press.

Davaras, C. (1971) 'Protominoikon Nekrotapheion Aghias Foteias Siteas'. *Athens Annals of Archaeology* 4, 392–397.

Davis, J.L. (1981) 'Minos and Dexithea: Crete and the Cyclades in the Later Bronze Age'. In J.L. Davis and J.F. Cherry (eds) *Papers in Cycladic Prehistory.* Los Angeles, UCLA Institute of Archaeology, Monograph no. 14, 143–157.

Davis, J.L. (1984) 'Cultural Innovation and the Minoan Thalassocracy at Aghia Irini, Keos'. In Hägg and Marinatos (eds), 159–166.

Davis, J.L. (1986) *Keos V. Aghia Irini Period V.* Mainz, Von Zabern.

Deacon, A.B. (1934) *Malekula: A Vanishing People in the New Hebrides.* London, Routledge.

DeLaet, S. (ed.) (1976) *Acculturation and Continuity in Atlantic Europe.* Ghent, Dissertationes Archaeologicae Gandenses 16.

de Lanfranchi, F. and Weiss, M.C. (1973) *La Civilisation des Corses: Les Origines*. Corsica, Ajaccio, Maison de la Culture de la Corse.

de Lanfranchi, F. and Weiss, M.C. (1977) *Araguina-Sennola*. (Archeologia Corsa 2). Ajaccio, Maison de la Culture de la Corse.

de Lumley, H., Gagniere, S., Barral, L. and Pascal, R. (1963) 'La Grotte du Vallonet, Roquebrun, Cap-Martin'. *Bulletin du Musée d'Anthropologie Préhistorique de Monaco* 10, 5–20.

de Lumley, H. (1967) 'Découverte d'habitat de l'Acheuléen Ancien dans des depots Mindéliens sur le Site de Terra Amata'. *Comptes Rendus des Séances Academiques des Sciences Paris*, Ser. D, 264, 801–804.

de Lumley, H. and de Lumley, M-A. (1971) 'Découverte de Restes Humains Anténéanderthaliens, datés du début du Riss à la Caune d'Arago (Tautavel, Pyrénées-Orientales)'. *Comptes Rendus des Séances Academiques des Sciences*. Paris, Ser. D, 272, 1739–1742.

Demurtas, L.M. and Demurtas, S. (1984) 'I Protonuraghi: Nuovi Dati Per L'Oristanese'. In Waldren *et al.* (eds), 629–669.

Diamond, J.M. (1989) 'Quaternary Megafaunal Extinctions: Variations on a Theme by Paganini'. *Journal of Archaeological Science* 16, 167–175.

Dickinson, O. (1986) 'Early Mycenean Greece and the Mediterranean'. In Marazzi *et al.* (eds), 271–276.

Dickinson, O. (1994) *The Aegean Bronze Age*. Cambridge University Press.

Doumas, C. (ed.) (1980) *Thera and the Aegean World I*. London, Thames & Hudson.

Doumas, C. (1982) 'The Minoan Thalassocracy and the Cyclades'. *Archaologischer Anzeiger* 5–14.

Efstratiou, N. (1985) *Ayios Petros: A Neolithic Site in the Northern Sporades*. Oxford, British Archaeological Reports (International Series) no. 241.

Ericson, J.E. and Earle, T.K. (eds) (1982) *Contexts for Prehistoric Exchange*. London, Academic Press.

Evans, A. (1921) *The Palace of Minos*. London, Macmillan.

Evans, J.D. (1953) 'The Prehistoric Culture Sequence in the Maltese Archipelago'. *Proceedings of the Prehistoric Society* 19, 41–94.

Evans, J.D. (1959) *Malta*. London, Thames & Hudson.

Evans, J.D. (1964) 'Excavations on the Neolithic Settlement at Knossos, 1957–60. Part 1'. *Annual of the British School of Archaeology at Athens* 59, 132–240.

Evans, J.D. (1968) 'Knossos Neolithic. Part 2: Summary and Conclusions'. *Annual of the British School of Archaeology at Athens* 63, 267–276.

Evans, J.D. (1971a) *The Prehistoric Antiquities of the Maltese Islands: A Survey*. London, Athlone Press.

Evans, J.D. (1971b) 'Neolithic Knossos: The Growth of a Settlement'. *Proceedings of the Prehistoric Society* 37, 95–117.

Evans, J.D. (1973) 'Islands as Laboratories for the Study of Culture Process'. In A.C. Renfrew (ed.) *The Explanation of Culture Change: Models in Prehistory*. London, Duckworth, 517–520.

Evans, J.D. (1977) 'Island Archaeology in the Mediterranean: Problems and Opportunities'. *World Archaeology* 9, 12–26.

Evans, J.D. (1984) 'Maltese Prehistory: A Reappraisal'. In Waldren *et al.* (eds), 489–497.

Evans, J.D. and Renfrew, A.C. (1968) *Excavations at Saliagos, near Antiparos*. London, British School of Archaeology at Athens, Supplementary Volume, no. 5.

Evett, D. (1973) 'A Preliminary Note on the Typology, Functional Variability and Trade of the Italian Neolithic'. *Origini* 7, 63–152.

Fadda, M.A. (1984) 'Il Nuraghe Monte Idda di Posada e la Ceramica a Pettine in Sardegna'. In Waldren *et al.* (eds), 671–702.

Farrington, I.S. (ed.) (1985) *Prehistoric Intensive Agriculture in the Tropics.* Oxford, British Archeological Reports (International Series) no. 232.

Faure, P. (1963) 'Cultes des Sommets et Cultes des Cavernes en Crète'. *Bulletin de Corréspondence Héllenique* 87, 493–508.

Faure, P. (1965) 'Récherches sur le Peuplement des Montagnes de Crète: Sites, Cavernes et Cultes'. *Bulletin de Corréspondence Héllenique* 89, 27–63.

Faure, P. (1967) 'Nouvelles Récherches sur Trois Sortes de Sanctuaires Crètois, I.' *Bulletin de Corréspondence Héllenique* 91, 114–150.

Faure, P. (1969) 'Sur Trois Sortes de Sanctuaires Crètois, II'. *Bulletin de Corréspondence Héllenique* 93, 174–213.

Fedele, F. (1980) *Nurahghi: La Misteriosa Civiltà dei Sardi.* Milan, Joca.

Ferrarese-Ceruti, M.-L. (1979) 'Ceramica Micenea in Sardegna'. *Rivista di Scienze Preistoriche* 34, 243–253.

Ferrarese-Ceruti, M-L., Vagnetti, L. and LoSchiavo, F. (1987) 'Minoici, Micenei e Ciprioti in Sardegna nella Seconda Meta del II Millennio a.c.' In Balmuth (ed.), 7–38.

Firth, R. (1961) *We, The Tikopia.* London, Allen & Unwin.

Florit-Piedrabuena, C. (1969) 'La Funcionalidad de las Taulas Revelada par un Viesjo Texto'. *Revista de Menorca* 111.

Foose, T.J. (1983) 'The Relevance of Captive Populations to the Conservation of Biotic Diversity'. In C.M. Schonewald-Cox, S.M. Chambers, B. MacBryde and W.C. Thomas (eds) *Genetics and Conservation. A Reference for Managing Wild Animal and Plant Populations.* Mento Park (Ca), Benjamin & Cummings, 374–401.

Fosberg, F.R. (ed.) (1963) *Man's Place in the Island Ecosystem.* Honolulu, Bishop Museum Press.

Francaviglia, V. and Piperno, M. (1987) 'La Répartition et la Provenance de l'Obsidienne Archéologique de la Grotta dell'Uzzo et de Monte Cofano (Sicile)'. *Révue de l'Archéométrie* 11, 31–39.

French, E.B. (1985) 'The Mycenean Spectrum'. In Malone and Stoddart (eds), 295–303.

Friedman, J. and Rowlands, M.J. (1977) 'Notes Towards an Epi-genetic Model for the Evolution of Civilisation'. In J. Friedman and M.J. Rowlands (eds) *The Evolution of Social Systems.* London, Duckworth, 201–276.

Gale, G.H. (ed.) (1991) *Bronze Age Trade in the Mediterranean.* Goteborg, Paul Astroms Forlag, Studies in Mediterranean Archaeology, no. 90.

Gale, N.H. (1980) 'Some Aspects of Lead and Silver Mining in the Aegean'. In Doumas (ed.), 160–195.

Gale, N.H. and Stos-Gale, Z.A. (1981) 'Cycladic Lead and Silver Metallurgy'. *Annual of the British School of Archaeology at Athens* 76, 169–224.

Gale, N.H. and Stos-Gale, Z.A. (1986) 'Oxhide Ingots in Crete and Cyprus and the Bronze Age Metals Trade'. *Annual of the British School of Archaeology at Athens* 81, 81–100.

Gallin, L. (1989) 'Architectural Attributes and Intersite Variation: A Case Study of the Sardinian Nuraghi', unpublished PhD thesis, University of California, Los Angeles.

Gasull, P., Lull, V. and Sanhauja, M.E. (1984) *Son Fornes I: La Fase Talayotica. Ensayo de Reconstruccion Socio-Economica de una Comunidad Prehistorica de la Isla de Mallorca.* Oxford, British Archaeological Reports (International Series) no. 209.

Gesell, G.C. (1987) 'The Minoan Palace and Public Cult'. In Hägg and Marinatos (eds), 123–128.

Giardiono, C. (1992) 'Nuraghic Sardinia and the Mediterranean: Metallurgy and Maritime Traffic'. In Tykot and Andrews (eds), 279–292.

Graham, J.W. (1959) 'The Residential Quarter of the Minoan Palace'. *American Journal of Archaeology* 63, 47–52.

Graham, J.W. (1977) 'Bathrooms and Lustral Basins'. In K.H. Kinzl (ed.) *Greece and the East Mediterranean in Ancient History and Prehistory. Studies Presented to Fritz Schachermeyer on the Occasion of his Eightieth Birthday.* Berlin, Linzel.

Gras, M. (1985) *Trafics Thyrreniens Archaiques.* Rome, Ecole Française de Rome.

Graziosi, P. (1953) 'Nuovi Graffiti della Grotta di Levanzo Egadi'. *Rivista di Scienze Prehistoriche* 8, 123–137.

Green, R.C. (1976) 'Lapita Sites in the Santa Cruz Group'. *Royal Society of New Zealand Bulletin* 11, 245–265.

Guilaine, J., Courtin, J., Roudil, J.-L. and Vernet, J.-L. (eds) (1987) *Premières Communautés Paysannes en Méditérannée Occidentale.* Paris, CNRS.

Hadjianastasiou, P. (1988) 'A Late Neolithic Settlement at Grotta, Naxos'. In E.B. French and K.A. Wardle (eds) *Problems in Greek Prehistory.* Bristol, Classical Press, 11–20.

Hägg, R. and Marinatos, N. (eds) (1984) *The Minoan Thalassocracy: Myth and Reality.* Stockholm, Acts of the Swedish Institute in Athens, no. 32.

Hägg, R. and Marinatos, N. (eds) (1987) *The Function of the Minoan Palaces.* Stockholm, Acts of the Swedish Institute in Athens, no. 35.

Hallager, E. (1987) 'A "Harvest Festival Room" in the Minoan Palaces? A Study of the Pillar-Crypt Area at Knossos'. In Hägg and Marinatos (eds), 169–177.

Hallam, B.R., Warren, S.E. and Renfrew, A.C. (1976) 'Obsidian in the West Mediterranean: Characterisation by Neutron Activation Analysis and Optical Emission Spectroscopy'. *Proceedings of the Prehistoric Society* 42, 85–110.

Halstead, P. (1981) 'Counting Sheep in Neolithic and Bronze Age Greece'. In I. Hodder, N. Hammond and G. Isaac (eds) *Pattern of the Past.* Cambridge University Press, 307–39.

Hankey, V. (1979) 'Crete, Cyprus and the South-East Mediterranean, 1400–1200 BC'. In *Acts of the International Archaeological Symposium 'The Relations Between Cyprus and Crete, c. 2000–500 BC'.* Nicosia, Department of Antiquities of Cyprus, 144–157.

Held, S.O. (1989) 'Colonisation Cycles on Cyprus I: The Biogeographic and Palaeontological Foundations of Early Prehistoric Settlement'. *Report of the Department of Antiquities of Cyprus* 1989, 7–28.

Helms, M.W. (1988) *Ulysses' Sail. An Ethnographic Odyssey of Power, Knowledge and Geographical Distance.* Princeton University Press.

Heltzer, M. (1988) 'Trade Relations between Ugarit and Crete'. *Minos* 23, 7–13.

Heltzer, M. (1989) 'The Trade of Crete and Cyprus with Syria and Mesopotamia, and their Eastern Tin Sources in the 18th–17th Centuries BC'. *Minos* 24, 7–27.

Heyerdahl, T. and Skolsvald, A. (1956) *Archaeological Evidence for Pre-Spanish Visits to the Galapagos Islands.* Boston, Society for American Archaeology, Memoir no. 12.

Hiller, S. (1984) 'Pax Minoica Versus Minoan Thalassocracy. Military Aspects of Minoan Culture'. In Hägg and Marinatos (eds), 27–31.

Hood, S. (1970) 'Tholos Tombs of the Aegean'. *Antiquity* 35, 168–170.

Hood, S. and Smyth, D. (1981) *Archaeological Survey of the Knossos Area.* London, British School of Archaeology at Athens, Supplement no. 14.

Hope-Simpson, R. and Dickinson, O.T.P.K. (1979) *A Gazeteer of Aegean Civilisation in the Bronze Age: Vol. I. The Mainland and Islands.* Goteborg, Paul Astroms Forlag, Studies in Mediterranean Archaeology, vol. 52.

Hughes, P.J. (1985) 'Prehistoric Man-induced Soil Erosion: Examples from Melanesia'. In Farrington (ed.), 398–408.

Iakovidis, S. (1979) 'Thera and Mycenean Greece'. *American Journal of Archaeology* 83, 100–110.

Jacobsen, T.W. (1969) 'Excavations at Porto Cheli and Vicinity: Preliminary Report 11. The Franchthi Cave 1969–1971'. *Hesperia* 38, 343–81.

Jacobsen, T.W. (1973) 'Excavations at the Franchthi Cave, 1969–1971'. *Hesperia* 42, 45–88, 253–83.

Jones, R.E. and Day, P.H. (1987) 'Late Bronze Age Aegean and Cypriot Type Pottery on Sardinia: Identification of Imports and Local Imitations by Physico-chemical Analysis'. In Balmuth (ed.), 257–269.

Juniper, B.E. (1984) 'The Natural Flora of Mallorca, *Myotragus* and its Possible Effects and the Coming of Man to the Balearics'. In Waldren *et al.* (eds), 145–163.

Kaplan, S. (1976) 'Ethnological and Biogeographical Significance of Pottery Sherds from Nissan Island, Papua New Guinea'. *Fieldiana: Anthropology* 66, no. 3.

Karageorghis, V. (1958) 'Origine du Syllabaire Chypro-Minoen'. *Révue Archéologique* 1–19.

Karageorghis, V. (1976) *Kition. Mycenean and Phoenician Discoveries on Cyprus.* London, Thames & Hudson.

Karageorghis, V. (1982) *Cyprus from the Stone Age to the Romans.* London, Thames & Hudson.

Kavvadias, G. (1984) *I Palaiolithiki Kephalonia: O Politismos tou Physkardou.* Athens, Phytraki.

Keegan, W.F. and Diamond, J.M. (1987) 'Colonisation of Islands by Humans: A Biogeographical Perspective'. In M.B. Schiffer (ed.) *Advances in Archaeological Method and Theory* 10, 49–92.

Keller, D.R. and Rupp, D.W. (eds) (1983) *Archaeological Survey in the Mediterranean.* Oxford, British Archaeological Reports (International Series) no. 155.

Kirch, P.V. (1984) *The Evolution of the Polynesian Chiefdoms.* Cambridge University Press.

Kirch, P.V. (1986) 'Exchange Systems and Inter-Island Conflict in the Transformation of an Island Society: The Tikopia Case'. In Kirch (cd.), 33–41.

Kirch, P.V. (ed.) (1986) *Island Societies: Archaeological Approaches to Evolution and Transformation.* Cambridge University Press.

Kirch, P.V. and Yen, D.E. (1982) *Tikopia: the Prehistory and Ecology of a Polynesian Outlier.* Honolulu, Bishop Museum Bulletin 238.

Klein-Hofmeijer, G., Sondaar, P.Y., Alderliesten, C., Van der Borg, K. and De Jong, A.F.M. (1987) 'Indications of Pleistocene Man on Sardinia'. *Nuclear Instrumental Methods in Physical Research* B29, 166–168.

Klemmer, K. (1959) 'Thunfisch Fang in Mittelmeer'. *Natur und Volk* 86, 173–179.

Knapp, A.B. (1990) 'Ethnicity, Entrepreneurship and Exchange: Mediterranean Inter-island Relationships in the Late Bronze Age'. *Annual of the British School of Archaeology at Athens* 85, 115–153.

Kopper, J. (1984) 'Canet Cave, Esporles, Mallorca'. In Waldren *et al.* (eds), 61–69.

Kopper, J. and Waldren, W. (1968) 'Balearic Prehistory, A New Perspective'. *Archaeology* 21, 108–115.

Kousoutflakis, G. (1990) 'Trade Mechanisms in the Early Bronze Age'. *Hydra* 7, 27–39.

Krzyszkowska, D. (1988) 'Ivory in the Aegean Bronze Age: Elephant Tusk or Hippopotamus Ivory?'. *Annual of the British School of Archaeology at Athens* 83, 209–234.

Krzyszkowska, D. and Nixon, L. (eds) (1983) *Minoan Society.* Bristol, Classical Press.

Laffineur, R. (1984) 'Myceneans at Thera: Further Evidence'. In Hägg and Marinatos (eds), 133–139.

Lagarce, J.E., Lagarce, A., Bounni, A. and Saliby, N. (1983) 'Les Fouilles à Ras Ibn Hani en Syrie (Campagnes de 1980, 1981 et 1982)'. *Académies des Inscriptions et Belle Lettres: Comptes Rendus Avril-Juin 1983.* 249–290.

Layard, J. (1942) *Stone Men of Malekula.* London, Chatto & Windus.

Leach, J.W. and Leach, E. (eds) (1983) *The Kula: New Perspectives on Massim Exchange.* Cambridge University Press.

Le Brun, A. (ed.) (1984) *Fouilles Récentes à Khirokitia (Chypre), 1977–1981.* Paris, Editions Récherche sur les Civilisations, ADPF, Memoire no. 41.

Leekley, D. and Noyes, R. (1975) *Archaeological Excavations in the Greek Islands.* Park Ridge Press, New Jersey.

Leisner, G. and Leisner, V. (1943) *Die Megalithgraber der Iberischen Halbinsel. I. Der Suden.* Berlin, Walter de Gruyter.

Lewthwaite, J. (1984a) 'Pastore, Padrone: The Social Dimensions of Pastoralism in Pre-Nuraghic Sardinia'. In Waldren *et al.* (eds), 251–268.

Lewthwaite, J. (1984b) 'The Neolithic of Corsica'. In C. Scarre (ed.) *Ancient France.* Edinburgh University Press, 146–181.

Lewthwaite, J. (1985a) 'From Precocity to Involution: The Neolithic of Corsica in its West Mediterranean Context'. *Oxford Journal of Archaeology* 4, 47–68.

Lewthwaite, J. (1985b) 'Colonialism and Nuraghismus'. In Malone and Stoddart (eds), 151–220.

Lewthwaite, J. (1990) 'Isolating the Residuals: the Mesolithic Basis of Man–Animal Relationships on the Mediterranean Islands'. In C. Bonsall (ed.) *The Mesolithic in Europe.* Edinburgh, John Donald, 541–555.

Lewis, D. (1972) *We, The Navigators.* Honolulu, University Press of Hawaii.

Lilliu, G. (1959a) 'The Nuraghi of Sardinia'. *Antiquity* 33, 32–38.

Lilliu, G. (1959b) 'The Protocastles of Sardinia'. *Scientific American* 201, 62–69.

Lilliu, G. (1960) 'Primi Scavi del Villaggio Talaiotico di Ses Paisses'. *Instituto Nazionale d'Archeologia e Storia* New Series, 9.

Lilliu, G. (1962a) 'Cenno Sui Piu Recenti Scavi del Villaggio Talaiotico di Ses Paisses ad Arta-Maiorca, Baleari'. *Archeologica Studi Sardo* 28, 63–87.

Lilliu, G. (1962b) *I Nuraghi: Torre Preistoriche di Sardegna.* Cagliari, La Zattera.

Lilliu, G. (1975) 'Prima dei Nuraghi' and 'Al Tempo dei Nuraghi'. In *La Societa in Sardegna nei Secoli.* Turin, Edizioni RAI, 7–31.

LoSchiavo, F. (1982) 'Copper Metallurgy in Sardinia During the Late Bronze Age: New Prospects on its Aegean Connections'. In J.D. Muhly, R. Maddin and V. Karageorghis (eds) *Early Metallurgy in Cyprus, 4000–500 BC.* Nicosia, Pierides Foundation, 271–282.

LoSchiavo, F. (1984) 'Appunti sull'Evoluzione Culturale della Sardegna nell'Eta dei Metalli'. *Nuovo Bulletino Archeologico Sardo* 1, 21–40.

LoSchiavo, F. (1988) 'Complesso Nuragico di Sa Sedda 'e Sos Carros'. In *Sardegna Centr-Orientale dal Neolitico all Fine del Mondo Antico.* Sassari, Dessi, 99–101.

LoSchiavo, F. (1989) 'Le Origini della Metallurgica ed il Problema della Metallurgica nella Cultura di Ozieri'. In L.D. Campus (ed.) *La Cultura di Ozieri: Problematiche e Nuove Acquisizioni. Atti del I Convegno di Studio (Ozieri 1986–7).* Ozieri, Il Torchietto, 272–292.

Luckerman, F. and Moody, J. (1985) 'The Wild Country West of Idha: The Prehistory of the Khania Nomos'. *American Journal of Archaeology* 89, 338–349.

MacArthur, N., Saunders, I. and Tweedie, R. (1976) 'Small Population Isolates: A Micro-Simulation Study'. *Journal of Polynesian Society* 85, 307–326.

MacArthur, R.H. and Wilson, E.O. (1967) *The Theory of Island Biogeography.* Princeton University Press.

MacGillivray, J.A. (1984) 'Cycladic Jars from Middle Minoan Contexts at Knossos'. In Hägg and Marinatos (eds), 153–158.

MacGillivray, J.A. (1987) 'Pottery Workshops of the Old Palace Period in Crete'. In Hägg and Marinatos (eds), 273–279.

Mackey, M. and Warren, S.E. (1983) 'The Identification of Obsidian Sources in the Monte Arci Region of Sardinia'. In A. Aspinall and S.E. Warren (eds), *Proceedings of the 22nd Symposium on Archeometry.* University of Bradford, 420–431.

Malinowski, B. (1922) *Argonauts of the Western Pacific.* London, Routledge.

Malone, C.A.T. (1985) 'Pots, Prestige and Ritual in Neolithic Southern Italy'. In Malone and Stoddart (eds) (vol. ii), 118–151.

Malone, C.A.T. and Stoddart, S.K.F. (eds) (1985) *Papers in Italian Archaeology IV.* Oxford, British Archaeological Reports (International Series) no. 265.

Marazzi, M., Tusa, S. and Vagnetti, L. (eds) (1986) *Traffici Micenei nel Mediterraneo.* Taranto, Istituto per la Storia e l'Archeologia della Magna Grecia.

Marcello, A., Pretti, S. and Salvadori, I. (1978) *Carta Metallurgenica della Sardegna.* Firenze, Servizio Geologico d'Italia.

Marinatos, N. (1984a) 'Minoan Threskeiocracy on Thera'. In Hägg and Marinatos (eds), 167–178.

Marinatos, N. (1984b) *Art and Religion in Thera.* Athens, Mathioulakis.

Marinatos, N. (1987) 'Public Festivals in the West Courts of the Palaces'. In Hägg and Marinatos (eds), 135–143.

Marinatos, S. (1939) 'The Volcanic Destruction of Minoan Crete'. *Antiquity* 13, 425–439.

Marthari, M. (1984) 'The Destruction of the Town at Akrotiri, Thera, at the beginning of Late Cycladic I. Definition and Chronology'. In MacGillivray and Barber (eds), 119–133.

Mascaro-Pasarius, J. (1958) *Els Monuments Megalitics a l'Illa de Menorca.* Barcelona, Mahon.

Mascaro-Pasarius, J. (1968) *Las Taulas.* Barcelona, Mahon.

Matsas, D. (1984) 'Mikro Vouni Samothrakis: Mia Proistorichi Koinotita s'ena visiotiko systima tou BA Aigaiou'. *Anthropologika* 6, 73–94.

Matsas, D. (1987) 'Samothraki 1987: Archaiologike kai Ethnoarchaiologikes Ergasisies'. *To Archaiologiko Ergo Sti Makedonia kai Thraki* 1, 499–503.

Mayr, E. (1954) 'Change of Genetic Environment and Evolution'. In J. Huxley, A.C. Hardy and E.B. Ford (eds) *Evolution as a Process.* London, Allen & Unwin, 157–80.

McCall, G.M. (1979) 'Kinship and Environment on Easter Island: Some Observations and Speculations'. *Mankind* 12, 119–137.

McEnroe, J. (1982) 'A Typology of Minoan Neopalatial Houses'. *American Journal of Archaeology* 86, 3–19.

Mead, M. (1943) *The Coming of Age in Samoa: A Study of Adolescence and Sex in Primitive Societies.* Harmondsworth, Penguin.

Meillassoux, C. (1964) *Anthropologie Economique des Gouro de Cote d'Ivoire.* Den Haag, Mouton.

Meillassoux, C. (1967) 'Récherche d'un Niveau de Détermination dans la Société Cynégétique'. *L'Homme et la Société* 6, 24–36.

Meillassoux, C. (1972) 'From Reproduction to Production'. *Economy and Society* 1, 93–105.

Melas, E.M. (1985) *The Islands of Karpathos, Saros and Kasos in the Neolithic and Bronze Age.* Goteborg, Paul Astroms Forlag, Studies in Mediterranean Archaeology, no. 68.

Mellaart, J. (1975) *The Neolithic of the Near East.* London, Thames & Hudson.

Métraux, A. (1957) *Easter Island: A Stone Age Civilisation in the Pacific.* London, André Deutsch.

Micha-Sarantea, E. (1980) Stoichea Lithotechnias tis Palaiolithikis Epochis sto nisi tis Euboias'. *Athens Annals of Archaeology* 11, 209–13.

Michels, J.W., Atzeni, E., Tsong, I.S.T. and Smith, G.A. (1984) 'Obsidian Hydration Dating in Sardinia'. In Balmuth and Rowland (eds), 83–113.

Milliken, S. and Skeates, R. (1989) 'The Almini Survey: The Mesolithic–Neolithic Transition in the Salerno Peninsula'. *Bulletin of the Institute of Archaeology* 26, 77–98.

Mirié, S. (1979) *Das Thronraumareal des Palastes von Knossos. Versuch einer Neuinterpretation seiner Entstehung und seiner Funktion.* Bonn, Saarbrucher Beitrage zur Altertumswissenschaft, 26.

Montague, S.P. (1980) 'Kula and Trobriand Cosmology'. *Journal of Anthropology* 2, 70–94.

Moody, J. (1983) 'Khania Archaeological Site Survey'. In Keller and Rupp (eds), 301–302.

Moody, J. (1987) 'The Minoan Palace as Prestige Artifact'. In Hägg and Marinatos (eds), 235–241.

Morgan, R. (1956) *World Sea Fisheries.* London, Methuen.

Mosimann, J.E. and Martin, P.S. (1975) 'Simulating Overkill by Palaeoindians'. *American Scientist* 63, 304–313.

Muhly, J.D., Maddin, R. and Stech, T. (1988) 'Cyprus, Crete and Sardinia: Copper Oxhide Ingots and the Bronze Age Metals Trade'. *Report of the Department of Antiquities of Cyprus* (1988), 281–298.

Munn, N. (1983) 'Gawan Kula: Spatiotemporal Control and the Symbolism of Influence'. In Leach and Leach (eds), 277–308.

Munn, N. (1986) *The Fame of Gawa.* Cambridge University Press.

Mylonas, G.E. (1959) *Aghios Kosmos: An Early Bronze Age Settlement and Cemetery in Attica.* Princeton University Press.

Negbi, O. (1992) 'Early Phoenician Presence in the Mediterranean Islands: A Reappraisal'. *Journal of Mediterranean Archaeology* 96, 599–615.

Nicholson, R.B. (1965) *The Pitcairners.* Sydney, Angus & Robertson.

Niemeier, W.-D. (1980) 'Die Katastrophe von Thera und die Spatminoische Chronologie'. *Jahrbuch des Deutschen Archaologischen Instituts* 95, 1–76.

Niemeier, W.-D. (1984) 'The End of the Minoan Thalassocracy'. In Hägg and Marinatos (eds), 205–215.

Niemeier, W-D. (1987) 'On the Function of the 'Throne Room' in the Palace at Knossos'. In Hägg and Marinatos (eds), 163–168.

Nixon, L. (1987) 'Neo-palatial Outlying Settlements and the Function of the Minoan Palaces'. In Hägg and Marinatos (eds), 96–99.

Nordfeldt, A. (1987) 'Residential Quarters and Lustral Basins'. In Hägg and Marinatos (eds), 187–194.

Nordquist, G.C. (1987) *A Middle Helladic Village: Asine in the Argolid.* University of Uppsala, BOREAS, vol.16.

Orsi, P. (1892) 'Necropoli Sicula presso Siracusa con Vasi e Bronzi Micenei'. *Monumenti Antichi Lincei* 2, 5–35.

Orsi, P. (1896) 'Thapsos'. *Monumenti Antichi Lincei* 6, 89–152.

Patton, M. (1993) *Statements in Stone: Monuments and Society in Neolithic Brittany.* London, Routledge.

Peatfield, A. (1987) 'Palace and Peak: The Political and Religious Relationship between Palaces and Peak Sanctuaries'. In Hägg and Marinatos (eds), 89–93.

Pericot-Garcia, L. (1972) *The Balearic Islands.* London, Thames & Hudson.

Perlès, C. (1979) 'Des Navigateurs Méditeranénens il y a 10,000 ans'. *La Récherche* 10, 82–83.

Perlès, C. (1992) 'Systems of Exchange and Organisation of Production in Neolithic Greece'. *Journal of Mediterranean Archaeology* 5, 115–164.

Phillips, P. (1975) *Early Farmers of West Mediterranean Europe.* London, Hutchinson.

Phillips, P. (1992) 'Western Mediterranean Obsidian Distribution and the European Neolithic'. In Tykot and Andrews (eds), 71–82.

Plantalamor, Ll. and Rita, M.C. (1984) 'Formas de Poblacion durante el Segundo y Primero Milenio BC en Menorca: Son Mercer de Baix, Transicion entre la Cultura Pretalayotica y Talayotica'. In Waldren *et al.* (eds), 797–826.

Popper, K. (1959) *The Logic of Scientific Discovery.* London, Hutchinson.

Poursat, J.-C. (1966) 'Une Sanctuaire du Minoen Moyen II à Mallia'. *Bulletin de Corréspondence Héllenique* 90, 514–551.

Poursat, J.-C. (1987) 'Town and Palace at Mallia in the Protopalatial Period'. In Hägg and Marinatos (eds), 75–76.

Puxeddu, C. (1958) 'Giacimenti di Ossidiana del Monte Arci in Sardegna e sua Irradiazione'. *Studi Sardi* 15, 10–66.

Re, L. (1986) 'Importazioni di Ceramica d'Uso Corrente sull'Isola di Vivara'. In Marazzi *et al.* (eds), 162–164.

Reille, M. (1977) 'Quelques Aspects de l'Activité Humaine en Corse durant le Subatlantique, et ses Conséquences sur la Végétation'. In H. Laville and J. Renault-Miskovsky (eds) *Approche Ecologique de l'Homme Fossile.* Paris, Supplément au Bulletin AFEQ no. 47, 329–342.

Rellini, V. (1923) 'Miniere e Fonderie d'Eta Nuragica in Sardegna'. *Bulletino di Paletnologia Italiana* 43, 58–67.

Renfrew, A.C. (1967) 'Colonialism and Megalithismus'. *Antiquity* 41, 276–88.

Renfrew, A.C. (1968) 'The Development and Chronology of Early Cycladic Figurines'. *American Journal of Archaeology* 73, 1–32.

Renfrew, A.C. (1972) *The Emergence of Civilisation: The Cyclades and the Aegean in the Third Millennium BC.* London, Methuen.

Renfrew, A.C. (1973) *Before Civilisation.* London, Cape.

Renfrew, A.C. (1974) 'Beyond a Subsistence Economy'. In C.B. Moore (ed.) *Reconstructing Complex Societies.* Supplement to the Bulletin of the American School of Prehistoric Research 78–93.

Renfrew, A.C. (1976) 'Megaliths, Territories and Populations'. In DeLaet (ed.), 198–220.

Renfrew, A.C. (1982) 'Alternative Models for Trade and Spatial Distribution'. In Ericson and Earle (eds), 71–90.

Renfrew, A.C. (1984) 'Trade as Action at a Distance'. In A.C. Renfrew (ed.) *Approaches to Social Archaeology.* Harvard University Press, 86–134.

Renfrew, A.C. (1985) *The Archaeology of Cult: The Sanctuary at Phylakopi.* London, Thames & Hudson.

Renfrew, A.C. and Level, E.V. (1979) 'Exploring Dominance: Predicting Polities from Centres'. In A.C. Renfrew and K.C. Cooke (eds) *Transformations: Mathematical Approaches to Culture Change.* New York, Academic Press, 145–167.

Renfrew, A.C. and Wagstaffe, M. (eds) (1982) *An Island Polity: The Archaeology of Exploitation in Melos.* Cambridge University Press.

Renfrew, A.C., Cann, J.R. and Dixon, J.E. (1965) 'Obsidian in the Aegean'. *Annual of the British School of Archaeology at Athens* 60, 225–247.

Rita, C. (1988) 'The Evolution of the Minorcan Pretalayotic Culture as Evidenced by the Sites of Morellet and Son Mercer de Baix'. *Proceedings of the Prehistoric Society* 54, 241–247.

Rossello-Bordoy, G. (1963) 'Cuevas Mallorquinas de Multiples Cameras'. *Studi Sardi* 18, 15–28.

Rowland, M.J. and Bent, S. (1982) 'Survey and Excavation on the Kedekede Hillfort, Lakeba Island, Lau Group, Fiji'. *Archaeology and Physical Anthropology in Oceania* 15, 20–50.

Rutkowski, B. (1986) *The Cult Places of the Aegean*. Yale University Press.

Rutter, J.B. (1985) 'An Exercise in Form as Function: The Significance of the Duck Vase'. *Temple University Aegean Symposium* 10, 16–41.

Rutter, J.B. and Zerner, C.W. (1984) 'Early Hellado-Minoan Contacts'. In Hägg and Marinatos (eds), 32–38.

Saflund, G. (1987) 'The Agoge of the Minoan Youth as Reflected by Palatial Iconography'. In Hägg and Marinatos (eds), 227–233.

Sahlins, M. (1955) 'Esoteric Efflorescence in Easter Island'. *American Anthropologist* 57, 1045–1052.

Sahlins, M. (1963) 'Poor Man, Rich Man, Big Man, Chief: Political Types in Melanesia and Polynesia'. *Comparative Studies in Society and History* 5, 285–303.

Sakellerakis, E. and Sakellerakis, Y. (1984) 'The Keftiu and the Minoan Thalassocracy'. In Hägg and Marinatos (eds), 197–203.

Sakellerakis, Y. (1977) 'Ta Kykladika Stoicheia ton Archanon'. *Athens Annals of Archaeology* 10, 93–115.

Sanders, E.A.C. and Reumer, J.W.F. (1984) 'The Influence of Prehistoric and Roman Migration on the Vertebrate Fauna of Menorca (Spain)'. In Waldren *et al.* (eds), 119–144.

Sanges, M. (1987) 'Gli Strati del Neolitico Antico e Medio nella Grotta Corbeddu di Oliena (Nuoru). Nota Preliminare'. *Atti della Reunione Scientifica dell'Instituto Italiano di Preistoria e Protostoria* 26, 825–30.

Savory, H.N. (1968) *Spain and Portugal*. London, Thames & Hudson.

Schachermeyer, F. (1980) 'Akrotiri: First Maritime Republic?'. In Doumas (ed.), 423–428.

Schofield, E. (1982) 'The Western Cyclades and Crete: A Special Relationship'. *Oxford Journal of Archaeology* 1, 9–25.

Scholte, B. (1979) 'From Silence to Discourse: The Structuraliste Impasse'. In S. Diamond (ed.) *Towards a Marxist Anthropology*. The Hague, Mouton.

Scoditti, G.M.G. and Leach, J.W. (1983) 'Kula on Kitava'. In Leach and Leach (eds), 249–273.

Serra-Belabre, M.L. (1964) *La Naveta d'Es Tudons, Mahon*. Madrid, Congreso Nacional de Arqueologia 10.

Shackleton, J.C., Van Andel, T.H. and Runnels, C.N. (1984) 'Coastal Palaeogeography of the Central and Western Mediterranean during the last 125,000 years, and its Archaeological Implications'. *Journal of Field Archaeology* 11, 307–14.

Shaw, J.W. (1987) 'A Palatial Stoa at Kommos'. In Hägg and Marinatos (eds), 101–110.

Sherratt, A.G. (1979) 'Plough and Pastoralism: Aspects of the Secondary Products Revolution'. In I. Hodder, N. Hammond and G. Isaac (eds) *Patterns in the Past*. Cambridge University Press, 261–305.

Simmons, A.H. (1991) 'Humans, Island Colonisation and Pleistocene Extinctions in the Mediterranean: The View from Akrotiri Aetokremnos, Cyprus'. *Antiquity* 65, 857–869.

Sondaar, P.Y. (1987) 'Pleistocene Man and Extinctions of Island Endemics'. *Memoires de la Société Géologique de France* 150, 159–65.

Sondaar, P.Y. and Spoor, C.F. (1982) 'Man and the Pleistocene Endemic Fauna of Sardinia'. In G. Giacobini (ed.) *Hominidae. Proceedings of the 2nd International Congress of Human Palaeontology, Turin*. Milan, Jaca Books.

Sondaar, P.Y., DeBoer, P.L., Sanges, M., Kotsakis, T. and Esu, D. (1984) 'First Report on a Palaeolithic Culture in Sardinia'. In Waldren *et al.* (eds), 29–59.

Sordinas, A. (1970) *Stone Implements from Northwestern Corfu, Greece*. Memphis State University Anthropological Research Centre, Occasional Paper no. 4.

Sotirakopoulou, P. (1989) 'The Earliest History of Akrotiri: The Late Neolithic and Early Bronze Age Phases'. In D. Hardy (ed.) *Thera and the Aegean World III*. vol. 2. London, Thera Foundation, 139–48.

Spriggs, M. (1985) 'Prehistoric Man-induced Landscape Enhancement in the Pacific: Examples and Implications'. In Farrington (ed.), 409–420.

Stevenson, C.M. (1986) 'The Socio-political Structure of the Southern Coastal Area of Easter Island: AD 1300–1864'. In Kirch (ed.), 69–77.

Stoddart, S., Bonanno, A., Gouder, T., Malone, C. and Trump, D. (1993) 'Cult in an Island Society: Prehistoric Malta in the Tarxien Period'. *Cambridge Archaeological Journal* 3, 3–19.

Stos-Gale, Z.A. and Gale, N.H. (1984) 'The Minoan Thalassocracy and the Aegean Metal Trade'. In Hägg and Marinatos (eds), 59–63.

Stos-Gale, Z.A. and Gale, N.H. (1992) 'New Light on the Provenance of the Copper Ingots Found on Sardinia'. In Tykot and Andrews (eds), 281–298.

Stringer, C., Howell, F. and Melentis, J. (1979) 'The Significance of the Fossil Hominid Skull from Petralona, Greece'. *Journal of Archaeological Science* 6, 235–253.

Tambiah, S.J. (1983) 'On Flying Witches and Flying Canoes: The Coding of Male and Female Values'. In Leach and Leach (eds), 171–200.

Taylor, D. (1975) *Some Locational Aspects of Middle Range Hierarchical Societies*. Ann Arbor (MI), University Microfilms.

Taylour, W. (1958) *Mycenean Pottery in Italy and Adjacent Areas*. Cambridge University Press.

Terrell, J. (1977) 'Human Biogeography in the Solomon Islands'. *Fieldiana: Anthropology* vol. 68, no. 1.

Terrell, J. (1986) *Prehistory in the Pacific Islands*. Cambridge University Press.

Theocaris, D.R. (1973) *Neolithic Greece*. Athens, National Bank of Greece.

Todd, I.A. (1987) *Vasilikos Valley Project 6: Excavations at Kalavassos-Tenta*. Goteborg, Paul Astroms Forlag Studies in Mediterranean Archaeology, no. 75.

Topp, C., Fernandez, J.H. and Plantalamor, Ll. (1976) 'Ca na Costa: A Megalithic Chamber Tomb on Formentera, Balearic Islands'. *Bulletin of the Institute of Archaeology* 13, 139–174.

Topp, C., Fernandez, J.H. and Plantalamor, Ll. (1979) 'Recent Archaeological Activities in Ibiza and Formentera'. *Bulletin of the Institute of Archaeology of the University of London* 19, 215–231.

Tore, G. (1984) 'Per una Rilettura del Complesso Nuragico di S'Urahi, Loc. Su Pardu, S. Vero Milis, Oristano (Sardegna)'. In Waldren *et al.* (eds), 703–723.

Torrence, R. (1986) *Production and Exchange of Stone Tools: Prehistoric Obsidian in the Aegean*. Cambridge University Press.

Trump, D.H. (1966) *Skorba*. London, Society of Antiquaries Research Report no. 22.

Trump, D.H. (1976) 'The Collapse of the Maltese Temples'. In G. de G. Sieveking, I.H. Longworth and K.E. Wilson (eds). *Problems in Economic and Social Archaeology*. London, Duckworth, 605–609.

Trump, D.H. (1980) *The Prehistory of the Mediterranean*. London, Allen Lane.

Trump, D.H. (1984) 'The Bonu Ighinu Project: Results and Prospects'. In Waldren *et al.* (eds), 511–522.

Tykot, R.H. (1992) 'The Sources and Distribution of Sardinian Obsidian'. In Tykot and Andrews (eds), 57–70.

Tykot, R.H. and Andrews, T.K. (eds) (1992) *Sardinia in the Mediterranean: A Footprint in the Sea.* Sheffield Academic Press.

Uberoi, J.P. Singh (1971) *The Politics of the Kula Ring.* Manchester University Press.

Vagnetti, L. (1983) 'Frammenti di Ceramica "Matt-Painted" Policroma da Filicudi Isole Eolie'. *Mélanges de l'Ecole Francaise de Rome* 95, 335–344.

Van Andel, T.H. and Runnels, C.N. (1988) 'An Essay on the Emergence of Civilisation in the Aegean World'. *Antiquity* 62, 234–247.

Vayda, A.P. and Rappaport, R.A. (1963) 'Island Cultures'. In Fosberg (ed.), 133–142.

Vermeule, E. (1964) *Greece in the Bronze Age.* Chicago University Press.

Vernet, J-L, Garcia, E.B., Almero, E.G. and Mora, T.R. (1984) 'Charcoal Analysis and the West Mediterranean Flora'. In Waldren *et al.* (eds), 165–172.

Vigne, J.D. (1987) 'L'Exploitation des Ressources Alimentaires Carnées en Corse du VII au IVe Millènaire'. In J. Guilaine, J. Courtin, J.-L. Roudil and J.-L. Vernet (eds) *Premières Communautés Paysannes en Méditeranée Occidentale.* Paris, CNRS, 193–199.

Vigne, J.D. (1989) 'Le Peuplement Préhistorique des Iles: Le Débat s'Ouvre en Sardaigne'. *Nouvelles de l'Archéologie* 35, 39–42.

Vigne, J.D., Marinval-Vigne, M.C., de Lanfranchi, F. and Weiss, M.-C. (1981) 'Consommation du 'Lapin-Rat' (Prolagus sardus Wagner) au Néolithique Ancien Méditerannéen. Abri d'Araguina-Sennola (Bonifaccio, Corse)'. *Bulletin de la Société Préhistorique Francaise* 78, 222–224.

Voza, G. (1972) 'Thapsos: Primi Resultati delle Piu Ricenti Riccerce'. *Atti della XIV Riunione Scientifica dell' Istituto Italiano di Preistoria e Protostoria (Puglia 1970),* 175–205.

Voza, G. (1973) 'Thapsos: Resoconto sulle Campagne di Scavo del 1970–71'. *Atti della XV Riunione Scientifica dell' Istituto Italiano di Preistoria e Protostoria (Verona-Trento 1972,* 133–157.

Waldren, W.H. (1982) *Balearic Prehistoric Ecology and Culture.* Oxford, British Archaeological Reports (International Series) S149.

Waldren, W.H., Chapman, R., Lewthwaite, J. and Kennard, R.-C. (eds) (1984) *The Deya Conference of Prehistory. Early Settlement in the West Mediterranean Islands and the Peripheral Areas.* Oxford, British Archaeological Reports (International Series) S229.

Warren, P. (1984) 'The Place of Crete in the Thalassocracy of Minos'. In Hägg and Marinatos (eds), 39–43.

Warren, P. (1987) 'The Genesis of the Minoan Palace'. In Hägg and Marinatos (eds), 45–56.

Watkins, T. (1981) 'The Chalcolithic Period in Cyprus: The Background to Current Research'. In J. Reade (ed.) *Chalcolithic Cyprus and Western Asia.* London, British Museum Occasional Paper 26, 9–20.

Watrous, L.V. (1982) *Lasithi. A History of Settlement on a Highland Plain in Crete.* Hesperia, Supplement no. 18.

Webster, G.S. (1991) 'Monuments, Mobilisation and Nuraghic Organisation'. *Antiquity* 65, 840–856.

Webster, G.S. and Michels, J.W. (1986) 'Palaeoeconomy in West Central Sardinia'. *Antiquity* 60, 226–229.

Weiner, A.B. (1976) *Women of Value, Men of Renown: New Perspectives in Trobriand Exchange.* Austin, Texas.

Whitehouse, R.D. (1972) 'The Rock-Cut Tombs of the Central Mediterranean'. *Antiquity* 46, 275–281.

Whitehouse, R.D. (1981) 'Megaliths of the Central Mediterranean'. In J.D. Evans, B.W. Cunliffe and A.C. Renfrew (eds) *Antiquity and Man. Essays in Honour of Glyn Daniel.* London, Thames & Hudson, 106–127.

Whitehouse, R.D. (1992) *Underground Religion: Cult and Culture in Prehistoric Italy.* London, Accordia Research Centre.

Whitelaw, T.M. (1983) 'The Settlement at Fournou Korifi, Myrtos, and Aspects of Early Minoan Social Organisation'. In Krzyszkowska and Nixon (eds), 324–345.

Whittle, A. (1985) *Neolithic Europe: A Survey.* Cambridge University Press.

Wiener, M. (1984) 'Crete and the Cyclades in Late Minoan I: The Tale of the Conical Cups'. In Hägg and Marinatos (eds), 17–26.

Wiener, M. (1990) 'The Isles of Crete? The Minoan Thalassocracy Revisited'. In D.A. Hardy (ed.) *Thera and the Aegean World. vol. I.* London, Thera Foundation, 128–160.

Williams, F.E. (1977) 'Trading Voyages from the Gulf of Papua'. In E. Schwimmer (ed.) *The Vailala Madness and Other Essays.* Honolulu, University Press of Hawaii, 48–72.

Williams-Thorpe, O., Warren, S.E. and Barfield, L.H. (1979) 'The Distribution and Sources of Archaeological Obsidian from Northern Italy'. *Preistorica Alpina* 15, 73–92.

Williamson, I. and Sabath, M.D. (1984) 'Small Population Instability and Island Settlement Patterns'. *Human Ecology* 12, 21–33.

Williamson, M. (1981) *Island Populations.* Oxford University Press.

Williamson, R.W. (1933) *Religious and Cosmic Beliefs of Central Polynesia. Vol. I.* New York, AMS Press.

Wylie, A. (1982) 'Epistemological Issues Raised by a Structuralist Archaeology'. In I. Hodder (ed.) *Structural and Symbolic Archaeology.* Cambridge University Press. 39–46.

Zammit, T. (1930) *Prehistoric Malta: The Tarxien Temples.* London, Mitford.

Zois, A. (1976) *Vasiliki I.* Athens Archaeological Society.

Zwicher, V., Virdis, P. and Ferrarese-Ceruti, M-L. (1980) 'Investigations on Copper Ore, Prehistoric Copper Slag and Copper Ingots from Sardinia'. In P.T. Craddock (ed.) *Scientific Studies in Early Mining and Extractive Metallurgy.* London, British Museum Occasional Paper no. 20, 135–164.

INDEX